SECRET SHADOWS *of* YESTERDAY

By

BRUCE STOCKDELL

ISBN: 0-75965-037-3

This book is printed on acid free paper.

1stBooks - rev. 10/1/01

CREDITS

Although the author is the only one who was involved in each of the incidents covered in this book it will quickly become evident that the work was not done alone. Much credit must go to the many people, scattered throughout the world, who contributed so greatly to the ultimate successes written about here.

Special note, and honor, goes to my dear wife, Imogene, who was not around when these events occurred and therefore made the perfect critical editor of the stories written as she was being exposed to them for the first time. At my request she become the "Devils Advocate" when she did not understand what was written or why. Reminding her that, at the end, the author must make the final decision but, until then, defend her position. Some very spirited, and even heated, discussions followed but it produced a better, and much more readable, document. A side benefit was that the lively discussions also resulted in our becoming much closer. Much credit goes to Imogene.

The Author

TABLE OF CONTENTS

ELECTRONICS

NUCLEAR

MISCELLANEOUS

PREFACE

This book is written by a career Air Force officer who started out as a "Buck Private" in US Army field artillery, made his way into flight training then persevered through three wars in a career which was anything but normal and most certainly not predictable. His course eventually led him into field collection for the Air Force Air Technical Intelligence Program and into some very unusual situations and predicaments, calling for innovative and unorthodox solutions. Fortunately his Commanders and/or supervisors extended an inordinate amount of trust and freedom of action which allowed him to proceed, on his own, and meet each situation as it arose.

One outstanding example, still vividly etched on his mind, involved Major General Harold Watson, the "father" of Air Technical Intelligence. He was "buried" while running a major intelligence collection operation requiring very tight security under the direct supervision of Gen Watson. The command channel was by telephone to the general's private number with instructions to call when a problem arose or when approval or directions were needed. A meeting would be arranged at some neutral hotel or restaurant. At these meetings, at the general's direction, it would be "Hal" and "Bruce", nothing military was to be mentioned and we would not go near nor mention any military base. A young Captain (two bars) meeting a Major General (two stars) on a personal basis— very heady stuff indeed.

On one occasion the General brought a full Colonel with him to impart, first hand, some information needed. The Colonel knew the author and, as he walked across the restaurant to join them, called out "Hello Captain Stockdell". As soon as we sat down the General proceeded to give the Colonel the most powerful "dressing down" ever heard. It was embarrassing but it was the kind of slip that could not be condoned for it could have blown our cover. Fortunately it did not this time.

At these meetings the problem (s) would be discussed including possible pitfalls, possible solutions and expected gains. Usually the field recommendations and the reasoning behind them would be presented first. If the General agreed, he would say "do it and I will back you". Such strong support encouraged harder work and led to some spectacular successes. There was, of course, constant awareness that such support would continue only as long as we were successful. Fortunately we were.

Over the years a number of bosses, fascinated by what was being done, asked if a book was going to be written about this work. At that time it did not appear likely because so many of the people recruited and trained were still working and many of the collection techniques and methods developed were still in use. Primary protection of the people and security of the projects depended on keeping our mouths shut. This has been done for these many years.

Now the world has changed along with some of the standards and the rules. Recent defections, traitors, and reported penetrations have given Intelligence operations such a black eye that it was determined, to balance the picture, our people deserved to be told some of our success stories. Hq USAF approval was requested and received. Through them clearance was granted from DOD and other senior US intelligence agencies to tell this story, with only two minor restrictions. These were that intelligence agents' names were not to be used and specific details of some equipment collected were not to be revealed. Living within these restrictions these stories are now presented. Every case presented is true and the author, personally, was involved in each. It is hoped you will find the stories both interesting and informative.

INTELLIGENCE

Intelligence is often referred to as the second oldest profession in the world. Yet in each generation it seems we must again relearn the same old lessons. In today's intelligence operations we have many types and classifications of collection systems, methods and mechanisms. These include, but are not limited to, COMINT-or Communications Intelligence; RADINT-or Radar Intelligence; SIGINT-or Signals Intelligence; and HUMINT-or Human Intelligence. Each of these programs, except HUMINT, requires and employs exotic, very complex and expensive high technology equipment. These programs, in turn, receive much support in promotion and funding at high levels of government and industry. Unfortunately, due to this popularity, the exotic systems are often supported at the expense of HUMINT, even though they cannot supply the type and quality of intelligence that HUMINT can provide. This was learned again, to our sorrow, in Iraq during Desert Storm, when the exotic systems could not provide the type and timeliness of intelligence, that can only be provided by human beings on the scene. The earliest example we have of such intelligence is in the Old Testament of the Bible when Moses sent Joshua, Caleb and ten others to spy out the Promised Land before the Israelites invaded. This human element, and the type of information that can only be provided by human sources, is the subject of this book.

HUMINT can go many directions and take many forms. The activities described here are sometimes those employing the facilities, the people and the opportunities of the international commercial market place.

Throughout history some of the best intelligence networks and sources have been those found in the marketplace. Tradesmen have access to people, information and sources that professional intelligence operatives can seldom hope to develop. For this reason, an operation was designed to tap these sources by placing professional intelligence operatives in world-wide commercial channels.

This book deals principally with that effort-the Foreign Material Program (FMP) of the United States Air Force, and its parts, the Foreign Material Acquisition Program (FMAP) and the Foreign Material Exploitation Program (FMEP). This program gave the US Air Force access to information on governmental trends, policies and decisions while also allowing procurement of items of both military and civilian equipment of great interest, which were not otherwise available. Sometimes equipment was obtained from the battle fields; sometimes purchased through commercial channels; while at still other times merely obtaining the specifications of items was sufficient. All of these efforts provided very good, factual, intelligence information which had direct application to advanced operational and R&D planning, giving the capabilities of

1

the systems which might be faced and the technological threats which had to be designed against. The author was deeply involved in the establishment and operation of this program from its conception and all cases discussed are from first hand knowledge.

To implement this program a new "BOOK" had to be written, the operatives assigned had to be carefully selected, deeply buried, fully supported and given a wide latitude of operation. The "BOOK" was written as the program unfolded.

Until now this story could not be told, as many of the techniques developed were still being used and many key people involved were still alive and active. Permission to print has now been received.

The initial impetus for starting this program came when the Soviet Union first decided, in the early 1950's, to display their wares outside the Soviet Bloc. For their first effort, in the early 1950's, they chose a location still behind the Iron Curtain but accessible to the west-the Vienna, Austria International Trade Fair. Already living and working there led to being assigned to cover the Fair for the United States. The equipment displayed at this first effort was of low level importance, but did give our scientists and engineers their first close look at the latest Soviet BLOC technology as represented in their heavy machine tools, carbide cutting tools, chemical-resistant plastic hose and other areas. More important, it proved we could satisfy many of our intelligence requirements through commercial channels. At later Fairs, through local business contacts, the entire Soviet exhibit was purchased. This provided additional proof that ours needs could be satisfied through commercial channels. To better accomplish this the Foreign Materiel program was started, aimed primarily at certain items of equipment, but always alert to any other collection opportunities.

Several locations throughout the world were opened at the same time including a cut-off operation established in a major US city.

Central control was maintained at Wright Patterson AFB, Ohio while the author was left in the field to run field collection operations.

Later Air Force operations were merged with Collection and Evaluation operations of the Army and Navy in Defense Intelligence Agency's Scientific and Technical Intelligence Division. At various times the author headed each of these offices. From the beginning this program was successful, as it provided greater direct identifiable dollar gain to the United States than it cost to run the program, but it was not without its growing pains and unusual experiences. I hope the readers will find these experiences both informative and enjoyable.

BACKGROUND

This brief personal background of the writer may help you better understand this story. It is a story written by one raised in a rural area of southern Indiana and It takes some very unusual, even bizarre, twists and turns that were totally unpredictable and unexpected. Included are adventure, excitement, hazardous escapes, joy, sorrow. Others just happened and cannot be explained or tagged.

While this is not intended to be an autobiography, the writer was the only one present, and personally involved, in every story, every operation, and every example presented so he is the only one who could write this book. No special knowledge nor special capability is claimed leading to the assignments and opportunities that made these happenings possible. The only attributes and characteristics that might are an unusual combination of education and experience coupled with a lifetime practice of accepting an assigned task, determining how to best get from A to B, then proceeding to complete the assignment without worrying about all the reasons it couldn't be done. This attitude, and the results produced, led to being assigned many very difficult, or even "impossible", tasks. It also resulted in the latter part of his career being spent almost exclusively in starting up new works for the Air Force. It produced a very challenging, and sometimes fascinating, career.

Raised in rural southern Indiana by very conservative, religious parents instilled principles and moral values that were, to a large degree, responsible for the directions that his life took and whatever successes achieved. The parents are gone now but are to be thanked for their lasting influence.

When a Sophomore in High School the writer decided that he wanted to be an engineer but knew that his parents, although supportive and very encouraging, could not help financially so, from that time, all efforts and all resources had to be dedicated to that goal. This caused foregoing a "normal" teen age life with many of it's pleasures such as dating, owning a car, and other things a teenager enjoys. Every penny had to be hoarded and the best possible grades obtained to prepare for engineering school. Although difficult it paid off with an Indiana State scholarship to Purdue University, his first choice, which helped greatly towards achieving his goals.

After High School he entered Purdue where he shared a room in a private home, and supported himself by drawing topographical maps for the US Department of Agriculture Extension Service and by waiting tables in a restaurant for his meals. At the time he carried a 22 credit-hour load in engineering. It was very hard but the first semester was successfully completed and preparation for the second started when he learned that the Selective Service Board would be calling him for active duty in the war before the end of that semester. So he did not return at that time.

Drafted into US Army field artillery he was assigned to basic training followed by Field Artillery Officer Candidate School (OCS). These assignments were probably based on his field artillery Reserve Officers Training Corps (ROTC) while at Purdue. Half way through basic training a notice was posted that the Air Corps Aviation Cadet training had opened up for any who wanted to apply. Since he had always wanted to be a pilot an application was submitted. Four applied and two, including the writer, were selected and immediately transferred to start Air Corps basic training. During Cadet classification testing after basic training he qualified for all three air crew positions; Pilot, Navigator and Bombardier. Of course Pilot training was chosen but the Army, in its "infinite wisdom", decided that Navigators were needed more than Pilots so off to Hondo, Texas for Navigator training.

All went well until about four weeks before graduation and receiving Navigators' wings when the author, along with three others, was called in and advised that the Air Corps no longer needed Navigators. Since all four had qualified for all three flying officer positions a choice was offered; remain and graduate as Navigators or transfer to Pilot or Bombardier training. All four transferred.

Flying schools in California and Texas provided many good, happy experiences and direct competition with some of the best students and athletes from some of the best Colleges and Universities in the United States. Survival required everything you had. It was a good experience.

Following pre-flight, Pilot Cadets were moved to primary flight training schools to learn how to fly as a pilot. Assigned to Jones Field, in Bonham, Texas a contract civilian flight school at Bonham, Texas, the hometown of Speaker of the House of Representatives Sam Rayburn he was further assigned to a civilian flight instructor. This instructor was one of the best, a barnstorming pilot who already had over 10,000 flying hours which was a tremendous amount for that time in history. He knew how to fly and how to teach it. Each instructor was assigned four students and the norm was that three would graduate. Getting off to a slow start, flying the PT-19 airplanes, all did not go well. The writer felt that he was number four and the future looked bleak until one fateful day. That day the priority maneuver was simulated emergency landings. In this exercise the instructor would unexpectedly cut the throttle back to idle and the cadet had to pick out the best open field in which to land, set up an approach pattern and then, without using power, fly the plane in for a simulated landing. At the last critical moment, before touch-down, the Instructor would decide whether or not a successful approach for landing had been made then apply power and the student would fly out. On this day, when the power was cut, the best field appeared to be directly under the airplane, the most difficult possible position. Immediately a double 360 degree overhead approach was set up and the plane brought in for a near perfect landing. At that point he jumped from fourth to first in the group and had no further problems in primary flight training.

Next was Basic Flight School at Perrin Field, in Sherman, Texas. Here was the much bigger and heavier BT-13 airplane. At this time instructors, who were all young military pilots, were being moved out to combat so fast that eight different instructors were assigned for the first eight flights. Since all were more interested in what was going to happen to them than they were in their students, very little actual instruction was received in those eight flights. Unfortunately, eight hours was also the trigger point for the first check ride. A senior check pilot was assigned and the ride was a disaster for very little instruction had been received in the maneuvers tested. On the ground, when he asked what the students problem was, the check pilot was told that there had been eight different instructors for eight rides but he had no one none assigned at that time. This triggered a meeting with the Squadron Commander, followed by second check ride with the Commander. On this ride the Commander would say what he wanted done then ask if instruction had been received on that maneuver. If so, it was to be performed. If not, the Commander instructed on how it was to be done. Upon landing he advised that the test was completed satisfactorily, then proceeded to straighten out the flight operations section to assure that no other cadets would be placed in such jeopardy. There was no further difficulty in Basic Flight training.

During the war years there probably was no stricter discipline than that in Aviation Cadet training. An example was at Basic Flight school. The group flew every day starting before daylight until after dark with classroom work and two hours of PT (physical training) fitted between flights. One day at the scheduled flight time the weather of rain and/or snow, low clouds and fog was so bad it was decided that it was too bad even for our flight to fly (it had to be really bad since we had flown in some very nasty weather since you can't always have perfect weather in combat). So, those in charge decided to indulge themselves in some disciplinary training. The flight was lined up in a brace and told to "reach for Texas". This meant standing at an exaggerated form of "Attention" with the shoulders pulled way back and the arms straining to touch the floor. At the same time all eyes had to be looking straight ahead and no part of the body could move. The group was kept in this position for nearly two hours before being released. Afterward the outline of every foot was visible on the floor from moisture which had been pulled up through the dry concrete by the warm shoes standing there so long. It was hard, but we survived.

Following Basic Flight Training the surviving Cadets were sent to the final phase, Advanced Flight School, some for single engine and the others for multi engine training. Next was Blackland Army Air Field at Waco, Texas for multi engine training. Selected to be one of the Flight Commanders over a part of the Cadet Corps led to more work, more responsibility, and sometimes some unpleasant duties but provided good training and everything has a price.

For example, one day two of the twin engined trainers collided directly over our barracks and all four pilots were members of his flight. Three of the four died instantly and the fourth was blown free of the airplane. When the survivor came to he said he was floating down in his parachute and did not know how he got out of the airplane nor when he opened his parachute. The debris, and parts of the bodies floated to the ground around our quarters and the flight had to clean up the remains of their buddies and roommates. It was a very deadly, and very sobering reminder of the seriousness of our mission.

The next day the flight marched, as usual, to the scheduled classes. When the civilian instructor noted three empty seats he immediately began to "chew out" the Flight Leader (the author) for people "not showing up for scheduled appointments". Finally he paused and asked for an explanation. The answer: "Sir, those are the three who didn't make it back yesterday". The instructor turned very white and needless to say the rest of the class period was very sober, serious and quiet. It was a hard day.

There was another accident in this phase of training. Everyone was under great pressure to build a cushion of flying hours so that interruptions caused by inclement weather or other things could be covered and the class still meet their requirements and graduate on schedule. The war waited for no one. Flights were scheduled to take off early in the morning hours so that no usable time would be lost. One morning, before being checked out, the writer was scheduled to take off before daylight with his instructor. Several planes took off without any problems. Then his instructor pushed the throttles forward, the plane went down the runway, lifted off normally and climbed to an altitude of about one hundred feet. At that point the left wing dropped and the instructor wasn't able to bring it back up. The normal corrective action for that condition is to turn the wheel clockwise and, simultaneously, push in on the right rudder pedal. Aware of the seriousness of the situation and thinking that, because of his short stature, the instructor might not be able to get full travel on the rudder the author physically checked both controls and found both to be at maximum travel. The plane was dropping rapidly with the wings in an almost vertical position when it hit the ground very hard, breaking off part of the left wing, violently throwing the plane over to break off part of the right wing, then going back and forth until it was finally brought under control in the middle of the triangle formed by the three runways. Other flights continued to successfully take off. In the darkness no one even knew there was trouble until we made a call on our radio to report the accident.

If the bank of the wings had been five degrees more we would have been vertical, probably the plane would have cart-wheeled and it is very possible that neither pilot would have survived.

Fortunately the instructor was at the controls so he had to meet the accident investigation board. Visible moisture (almost fog) was in the air at the time of the takeoff, just before dawn, and the Accident Board finding was that most likely frost had formed on the wings during takeoff and climb, stalling out the left wing. Why it didn't affect those taking off before, and after, could not be explained. We were fortunate.

Finally graduation day came and the coveted pilot wings were received.

Following graduation, no time was allowed for leave and all sent to combat crew training to finish preparation for combat. The author was sent to Yuma, AZ for B-17 Flying Fortress training. Arriving on the afternoon of the day scheduled it was learned that someone had "goofed" and two groups had arrived on the same day, but there was room for only one. Since the other group had arrived earlier and were already settled our group was given unexpected leave. His closest friend boarded a train going west as he boarded one going east. Both were derailed that day. It was wartime and travel could be hazardous to your health.

Trained in the B-17 Flying Fortress continued until the crews were no longer needed. Next was assignment to train in the Martin B-26 Marauder light bomber also at Yuma. Yuma was a combined center for checking out pilots, training aerial gunners and assembling combat crews so nearly every flight had multiple-mission assignments. Two of us, both former B-17 pilots, were assigned for our first flight in the B-26 with an instructor. Also aboard was a full complement of gunnery trainees, a full load of ammunition and full gas tanks for a maximum length mission. The aircraft was very heavy.

The plan was for the other pilot trainee to assist the instructor on take off and first half of the mission while the author observed from the Navigator's seat just behind the two pilots. Halfway through the mission we would switch places for the second half including the landing. As the instructor pushed the throttles forward to gain speed for takeoff he moved his hand from the throttles to the wheel to help pull the nose wheel off of the ground. When he did so he had his right hand in a "thumbs up position". This is the universal signal to retract the landing gear so that is what his co-pilot did. This was the first airplane with tricycle landing gear that both student pilots had flown. Until then only "tail draggers" (planes with a tail wheel) had been flown and they have completely different takeoff and landing characteristics. When he pulled the gear, unfortunately, the speed was only seventy five miles per hour down the runway in an airplane with a minimum flight speed of one hundred twenty plus miles per hour at that loading. The results were catastrophic. The instructor already had the nose gear off the runway so it snapped up immediately. Next in the normal cycle was retraction of the left main landing gear. As it started to go up the left wing dropped and the big propeller started biting into the runway making a terrific racket and pulling the airplane into a left turn. This airplane was equipped with the most powerful gasoline engine made at that time turning the largest four-bladed propeller in use. When the right gear folded the plane settled to the runway and slid up to a fence. It stopped just short of the fence which surrounded the ammunition storage depot for the Base.

The flight deck of the B-26 contained two pilot seats up front and, just behind them, seats for a navigator and a radio operator. There were two overhead hatches, one over the pilots and the other for the navigator to use when he took in-flight star readings with his octant. As the plane was sliding on its' belly the author jumped up and tried to open the Navigators hatch as a means of escape but the twisting of the fuselage kept it from opening. As the plane came to a stop the two Pilots managed to get their hatch open and all three climbed through it, ran to the tip of the left wing and dropped to the ground. The men in back were able to get out through the side door without difficulty. No one was hurt.

The noise the propellers made as they hit the runway had to have been horrendous for, by the time we three dropped to the ground, the Base Commander, the Hospital Commander, the Chief Flight Surgeon and even the Base Chaplain were there to greet us. As the three of us were talking to them it suddenly felt as if someone had hit me behind the knees with a ball bat. Whatever it was must have shown in my face because the Flight Surgeon immediately said, "sit down at once". He examined me and found nothing wrong, then said it was a case of delayed shock.

After a few moments we started examining the airplane. One of the first things noticed was that a one foot long piece of the left propeller blade had broken off and gone all the way through the fuselage of the plane, just about two inches below where the author was standing when trying to open the navigator's hatch. He had come within two inches of losing one, or both, of his feet. Someone was looking over his shoulder that day.

Shortly afterwards a transfer came through to a pilot pool where he flew the North American B-25 Mitchell light bomber.

At this point the author was selected for a very highly classified flying project but was not told the nature of that project. The transfer was immediate and, upon arriving in Roswell, NM learned that he had been selected to pilot a Boeing B-29 Super Fortress in the "A"bomb Group. At twenty years of age he was the oldest man in the crew of ten men. Later the crew was advised that the next station would be Tinian Island which turned out to be the launch point for the two nuclear weapons used against the Japanese. The war ended before we finished our training and jokingly we said "the Japanese heard we were coming so they quit".

At the end of the war the pilots were not discharged but released from active duty and placed in the Reserve Force. When the Korean war started this group was among the first recalled and the author was assigned as Pilot of a Boeing B-50 Lead crew in the number one bomb wing of the Strategic Air Force. From there he was selected to go into Scientific and Technical Intelligence work for the next twenty three years.

Entering Air technical Intelligence work at Wright-Patterson AFB, Ohio a crash course in the German language and training in Intelligence Field Collection techniques and methods was given before he was sent for duty behind the Iron Curtain in Vienna, Austria. This, in a nutshell, covers his life up to the start of the information presented in this book. Here the story about collection efforts starts.

THE STORY

The story starts with assignment, as a young US Air Force Lieutenant, to take the wife and five children to live and work behind the Iron Curtain at the time when Cold War tensions were at their highest. On the surface the assignment was as Pilot for the US High Commissioner for Austria and as Scientific and Technical Advisor to the United States High Commissioners Office (Embassy) in Vienna, Austria. At that time Vienna was one hundred miles behind the Iron Curtain in much the same physical position as Berlin. However, that is where the similarity ended. Berlin was constantly in the spotlight, the lines between the occupation zones were sharply drawn, and the demarcation lines were almost always fenced and guarded. Travel there, especially between zones, was very difficult, sometimes even impossible. By contrast, only the people living in Vienna knew where the occupation zone lines were and anyone could travel freely in all the zones. Vienna received very little attention and living there was very relaxed, compared to Berlin. The quarters provided were those confiscated from former top Nazi officials. Needless to say they were very good. With a large family a four story villa surrounded by a wrought iron fence and electrically controlled gate was assigned. No other non-Austrian family lived on the block. The US Army maintained a Commissary, Post Exchange and Movie Theater in the city that provided the essentials. It was very nice living.

Because of this relaxed atmosphere, it almost seemed as though there was a gentlemen's agreement between the powers that intelligence operations could be run through Vienna, both ways, and each side would look the other way or just track the other's operations without trying to obstruct them. While the author did fly the High Commissioner and advise the Embassy his primary assignment was to establish and run Scientific and Technical Intelligence collection efforts throughout the Iron Curtain countries. The relaxed atmosphere made this job easier and much more interesting for Vienna was the ideal spot for this work.

Scientific and Technical Intelligence collection operators are much different than regular intelligence collectors in that each must be a Scientist or Engineer, in selective fields, and must have a specified number of years industrial or military experience in their field. There were many good intelligence collectors who could go into the field, find, and count enemy radars, tanks, and other such equipment. These men did a good job, and they were needed, but they could not do the job the S&T collectors were trained to do. For example, S&T collectors could go into the field, find and observe a radar then come back with the frequency of operation, range, azimuth, height finding capabilities power requirements and other specifications possibly even including any apparent vulnerabilities that the equipment might have. It was sometimes said that there were a thousand men who could count for each one who could come back with this detailed information. Working with these men was very interesting.

The differing technical backgrounds and differing work experience made each field agent a distinct "personality" unto himself, so each directed his efforts in his own personalized way. As might be expected, this story reflects the author's own direction, and degree of effort.

Living arrangements in Vienna were very unusual. The accommodations were excellent but caused the family to be isolated from other American families since the closest Americans were several blocks away. This required becoming better acquainted with the Austrian people and to concentrating harder on learning the German language. The family grew to love the Austrian people. The other side of that coin was we were greatly exposed. This fact was graphically brought home one night when the Soviet NKVD (later the KGB) kidnapped a Czech refugee from his apartment just a block from our home. Although it wasn't needed, this was a grim reminder of why we were there and how vulnerable we were.

Early during the stay in Vienna it was learned that the Soviet Union had decided to mount their first effort, since World War II, to market their products in the west. For this they chose to participate in the annual International Trade Fair in Vienna, Austria. Undoubtedly contributing to this decision was the fact that, technically, Vienna was behind the Iron Curtain but, at the same time, accessible to the west. Western countries were not well informed on Soviet technologies so this was rated as a very important Intelligence opportunity. A meeting was convened, in Vienna, involving approximately twenty Western intelligence agencies to discuss the potential gain and how to conduct collection operations to insure complete coordinated coverage. Since he was the only US Scientific & Technical Intelligence collector present the author was assigned the task of covering their equipment and technology. To do this he teamed up with a British agent who could call on the Royal Army DDI Tech (their S&T intelligence organization) for assistance. It proved to be a very fruitful alliance for both sides.

Together, we decided that the best collection effort would be getting photographs, brochures and specifications documents, plus personal observation by qualified people. We had plenty of photographic equipment and a photo lab available but didn't have enough people to do the job properly, whereas the British friend said they had plenty of people but not enough equipment. The obvious solution was that the British provide the manpower, the Americans equip and train them, and we share the take. This was done to our mutual benefit. Press credentials for some of our photographers were also obtained while the others attended as tourists.

The plan of operation called for the two leaders to go through the exhibits and decide which items to target, trying not to get too greedy for any one day.

BRUCE STOCKDELL

Even so, some of our people were thrown out of an exhibit when they tried to get too close to an item while attempting to collect more detail. When that happened, we would assign another man to go into that exhibit the next day and finish the job. The British had provided twenty six men so the coordination job was complicated but it was done and complete coverage of all targets in the Soviet and Satellite exhibits was accomplished.

Late in the day each man dropped his exposed film at a designated point (changed each day) and a courier rushed the film to the photo lab for development. At the end of the day the two leaders would retire to the photo lab and study the negatives to determine the quality of coverage and which targets would have to be redone and which could be printed. The following day the technician would print two sets of the selected negatives, one for the British and one for the United States. A senior photo technician was offered on temporary loan because of the large volume of work in the lab. The offer was gratefully accepted, to our sorrow. This man decided to use a new experimental method to process our film. The result was the destruction of one entire days work and he was on the next train out of town. We had to repeat that day's effort.

Although this overall effort was a difficult job calling for long hours, it did have its lighter, and brighter, moments. The Soviet exhibit included a coil of chemical resistant plastic hose measuring about 15 mm in diameter. We were interested but were not able to get any information on the hose. Then, a British Captain offered to get samples for us. He said his wife had a very large handbag and a pair of very large shears (scissors) and they would come to the exhibit wearing long droopy coats and, while leaning over the exhibit, she would open the handbag allowing him to take out the shears, clip off a length of the hose, drop both shears and hose into the bag then walk away. All went as planned until, just as he clipped the hose, a heavy hand fell on his shoulder. He thought they were caught. It was one of his fellow officers playing a trick on him. Its a wonder that we didn't have a couple of cardiac arrests.

The Soviets also had an extensive display of carbide tipped cutting tools at the Fair. Since this was our first opportunity to see and learn of their capabilities in this field we were very interested in getting samples to study and analyze. But how? At this point a British Royal Army Major volunteered to get the desired samples. He was about 5' 4" tall, straight as a ramrod, and a perfect replica of the US movie idea of what a British Officer should look like. Since the weather was cold, he wore a uniform with a "greatcoat" as he entered the Soviet exhibit and proceeded to surreptitiously palm selected cutting tools dropping them into his greatcoat pockets. When he had what he wanted he started to leave the exhibit but his coat was so heavy that his legs almost bowed, but he marched out with his figure erect and his head held high. Later, we estimated that he had over fifty pounds of carefully selected cutting tools. What a performance! The British were fun to work with. Another agent managed to come up with the complete

12

detailed specifications of the most sophisticated machine tool shown. Altogether, it was a most profitable day.

When the fruits of the joint effort were assembled there were 800 high quality photographs, many brochures and many good samples for study and analysis. Complete reports were forwarded through the respective channels and both leaders received high praise for the operation and it's results. For several years afterwards this effort was held up as a model for the proper way to conduct such a collection operation.

Equally important, it provided proof that many of our collection requirements could be satisfied by getting, and analyzing, selected pieces of equipment. That thinking led, at the next trade fair, to buy the entire Soviet exhibit through a commercial cut-out. It was also easier and cheaper.

This success, and the satisfaction with the information gained, led the Air Force to set up the FMP (Foreign Materiel Program). This, in turn, led to some of agents being placed in the World's commercial trade channels. The balance of this book describes activities in that program as specific, planned programs against carefully selected targets are presented.

AVIATION

BRUCE STOCKDELL

AIRCRAFT

Since the Foreign Material Program written about here was first conceived as an Air Force program, it was only natural that much of the initial interest, and effort, would be directed toward aircraft and items associated with aircraft. It was also true during the Cold War that information in these areas was among the most closely guarded of the opposition's secrets and therefore would be very difficult to obtain. This only increased the challenge as it was decided to explore all possible channels and expend every possible effort to get possession of aircraft, parts, associated equipment and the specifications from whatever sources available. The sources would include, but were not limited to, battlefield pickup, commercial channels and aircraft which crashed and/or became available anywhere in the world. No opportunity was overlooked or ignored.

Teams were formed to go anywhere in the world on very short notice to evaluate or exploit all targeted aircraft and other pieces of equipment that became available. To man the teams, each Air force Research and Development (R&D) Installation and Test Center was tasked to identify and make available their best people in all of the appropriate technical fields.

The preferred goal was gaining control over the assigned target (s) and bringing it (them) to the United States for better and more complete exploitation. When this happened, the exploitation results were shared with the country from which the item was obtained.

The Flight Test Team included some of the most famous test pilots of that generation. If mentioned, all of their names would be familiar. The team members had to have current passports, current world-wide inoculations and be prepared to be out of the country within six hours to go anywhere in the world. All team members were very intrigued with the program and enthusiastically participated in all phases except for the inoculations, which were updated every six months. The biggest disappointment was that they weren't used often enough.

A classic example of the value of these teams was the defection of a pilot of a "Client State" in a Soviet made plane to a nearby country asking for asylum. The receiving country notified the US and asked for help. A team was dispatched and on arrival assessed all pertinent factors. It was obvious that the geographic proximity and political realities dictated that the plane would eventually have to be returned. It was also obvious that these factors coupled with the lack of local facilities precluded satisfactory exploitation on site.

The host country was asked to stall the return as long as possible while our team disassembled the plane, hauled it to the US, reassembled it and thoroughly exploited it including test flying. Then the process was reversed, the plane returned to the host country, then reassembled for return to its owner. All of this was done in fourteen days. These teams were good.

In addition to the teams we had other highly qualified test pilots and individuals who assisted very much in other individual ways. An outstanding example is Gen "Chuck" Yeager. He flew the Russian MIG-15, MIG-17, MIG-19, The Chinese MIG-19 the Russian MIG-21, MIG-24, MIG-27 and the MIG-29 fighter aircraft among others. Flying a sequence of aircraft from one manufacturer permits comparative evaluation and enables more accurate forecasting of future improvements. This, in turn, allows better planning in our US R&D&D and Flight Operations.

The following are some of the success stories and some of the failures in our collection efforts.

POLISH GLIDER

In the earlier days, while Europe was first starting to recover from World War II and the different countries' economies were in disarray, there was very little activity in the field of aeronautics which might interest the United States except in the Soviet Union. However one item in a Satellite country did draw attention. For two or three years in a row the Polish national glider team won several International glider competitions. No one was allowed to examine their glider but it appeared they had incorporated some new, or innovative, aerodynamic characteristics which gave them a competitive advantage. Since they would not allow anyone to examine their craft it was decided to try to buy one so a complete examination and test flight could be made. When a direct frontal approach was unsuccessful it was decided to try a "back door" approach. Commercial channels for intelligence acquisition had already been established so this was assigned as a test case.

It did not take long to find a commercial "Source" who promised to deliver one of the desired gliders at a given time, to a specified delivery point and at a firm price. An order was placed with the stipulation that only one particular model was wanted. The unit was to be made available for inspection prior to delivery and it would be accepted and paid for only if it was the one ordered. Anything else would be rejected as it would be in any other commercial transaction.

Since this procurement was "back door" the seller had specified that payment must be in used small bills of a "hard" currency that were not consecutively numbered. The currency could be German Marks, French Francs, English Pounds or US Dollars, with US Dollars preferred.

At this time in history, it was illegal (by US rules) for anyone, including Americans, to have greenbacks (US Dollars) in that part of the world. Printed script was the medium of exchange and new script was issued, without warning and on random schedules, to help combat the black markets which were rampant behind the Iron Curtain.

This complicated the operation but it was decided to pay in dollars for two reasons. First, it was the medium preferred by the seller, which would improve our standing as a customer for any future transactions and, second, it was easier for us to assemble in the form specified.

The purchase price of $25,000 was, at that time, a large sum of money when you reduce it to used twenty dollar bills. The bills filled a good sized briefcase that was quite heavy to carry and became an even heavier personal responsibility.

All went well until the glider was made available for inspection. At that point it was learned that the seller, too, had been not able to obtain the desired glider and was trying to palm off another model at a lower price. He was reminded that the order stipulated interest in only the one model. This refusal helped mark the buyers as hard-nosed commercial people who would accept only what was ordered.

It helped establish the operation but, at the same time, created a major problem for the buyer-a briefcase filled with a large sum of currency, in small bills, while traveling alone behind the Iron Curtain with no safe place to secure it. He could not go to any Embassy because there would be too many questions about who he was, what he was doing in "their" country and what he was doing with so much cash money in greenbacks. (It was often said that it would have been nice to have the State Department on our side!!!!). He also could not go to any bank since there would have even more questions. The only alternative was to make his way, cautiously, home and hold the cash there until it could safely be returned through his own secure channels.

This also presented problems. Not knowing how many people knew, or suspected, that he had the money in his possession everyone became suspect. Arriving home the bag of money was placed on the floor by the bed with a loaded pistol inside on top of the money and another loaded pistol under the pillow. Sleep was not very sound. The biggest fear was that one of the children would come into the bedroom in the middle of the night, startle him, and cause a catastrophic reaction before he was fully awake. Fortunately this did not happen.

This was the first major commercial effort and, although it was not completed to satisfaction, much was learned which proved valuable in later operations. The old saying "nothing ventured, nothing gained" proved to be true.

RINGNECK

The first opportunity to test the newly established commercial procurement system in a normal operation came with the chance to purchase a small Soviet general purpose helicopter direct from the factory. Although this helicopter, in itself, was of minimal interest it's usefulness in trying the newly established, and untested, commercial procurement channel made it, potentially, very valuable indeed. A decision was made to proceed and the order placed directly with the factory through the overseas part of our organization.

Starting a new classified project such as this requires a tremendous amount of planning, staffing and backup work. These were the responsibility of the author. Among other things are the assignment of an unclassified nickname. The name must be neutral so it can be used openly in an unclassified manner and not give a clue as to the true nature of the project. This project began to move on a Friday afternoon which meant Saturday was a work day because the pieces had to be quickly put in place. Unfortunately that was also the first day of pheasant hunting season. The Air Force Base where the office was located had worked hard to build a large pheasant population in their open spaces. They had succeeded to the point that the area was fast becoming overpopulated and a controlled hunt was scheduled to reduce the numbers. Those desiring to hunt could submit an application and a limited number of names were drawn. Those fortunate ones drawn were assigned selected areas, each of which almost assured a good bag. With a winning hand and a choice hunting area assigned for that Saturday morning a pleasant day had been planned. It was not to be. While sitting that morning trying to come up with a good nickname but thinking of the good hunting being missed and, at the same time it suddenly hit, "Project Ringneck" (for pheasants, of course)-so that was the name assigned. Later there was criticism because the word ringneck could imply the whirling main rotor blade of a helicopter. When it was explained all opposition disappeared and it remained PROJECT RINGNECK.

Next came the task of setting up the procurement channel to provide safe delivery of the helicopter to the desired destination in a manner that would assure security for all parties involved. The channels established were for a long term operation for it was hoped that much of it could be reused for follow-on operations.

To provide logical and acceptable cover a well known, and respected, US Aircraft Company was asked for their cooperation and support. The approach was directly to the president of the company. This company was selected because their product line fit and their president was a retired US Air Force Major General. Our plan was outlined and he asked how he and his company could be of assistance. It was explained that a good, logical, cover was needed under which to ship this helicopter in international trade channels without attracting undue attention. The desire was to use their corporate name for that purpose. After some discussion, he gave his permission for their company name to be painted in large letters on both the helicopter and on all the shipping crates. He also stated he would be his company's "project officer". To insure security he stated he would issue instructions to the company's fourteen divisions that any questions on this "project" were to be forwarded directly to him.

Before parting he asked if we did this kind of thing all the time. When assured that we did, he pulled a piece of paper out of his desk and asked if we could help him get the items on his list. It contained ten items that comprised a very good shopping list. He was informed we would try and we did but, sorry to say, we were never able to satisfy any of his requirements. This obvious involvement of large industry in commercial, or industrial, intelligence gave proof, if we needed any, that we were on the right course in our efforts to use commercial channels to gain intelligence information.

The company provided everything needed, and more. This first exposure to US industry on such matters set the pattern for many later projects. US industries were very helpful and most cooperative. They deserve a big vote of thanks from all of us. Without them, many of the accomplishments recorded in this book would not have been possible.

The procurement plan worked so well that the shipment arrived at a major US port, passed through and was shipped to the "cut-out" company in another location without our being aware, even though we were expecting it. it was then transferred to a flight test organization at a large US Air Force Base for test and evaluation. Carrying the Aircraft Company's name, it completed the entire test program, in the open, with only one person, as far as we know, recognizing it for what it was. That person was Gen Charles "Chuck" Yeager, the renowned test pilot, who was the first man to fly faster than the speed of sound. According to the report received, he went to the Commanding General's office and asked, "General, what is that Russian helicopter doing flying out there?" The General reportedly replied, "Chuck, you saw it, but you didn't." That was the closest we came to a security break on our first major procurement.

Later, having the privilege of knowing and working with Gen Yeager on several occasions we gained even greater respect for him and the author considers knowing him a major fringe benefit of his military career.

While, as stated earlier, the value of the small helicopter was not considered great we did gain enough knowledge to make it, by itself, worth the cost of the operation. The amazingly smooth success of our first operation represented our greatest gain as it generated enough confidence in our procurement mechanism to allow our proceeding on to greater and more intricate projects.

PROJECT LONG EARS

Good substantive requirements were constantly solicited from all intelligence branches of our government as these were needed to improve our quantity and quality of intelligence, open new technical fields in our procurement channels and increase the volume of business which, in turn, would help establish us in a stronger position in the commercial world. It was also necessary to realize the maximum return from our investment in the program. This resulted in extensive involvement in trade for electronics, planes, munitions, nuclear items and products from other scientific fields.

After the successful completion of Project Ringneck the next opportunity in aircraft turned out far better than we could have hoped. It was learned that a mid-size Russian helicopter could be purchased, again from the factory. This was of great interest because this helicopter, code named "Hound" by NATO, was an aircraft widely used in both land and sea based military versions, but also in many civilian applications.

When this opportunity was presented, top officials of the Army, Navy and the Air Force were very enthusiastic and urged that it be pursued with all possible urgency. They agreed to equally split the project cost of $1,800,000 which covered cost of the aircraft plus fifty percent (in dollar amount) spare parts-which is standard practice, world-wide, in the purchase of aircraft new from the factory including costs connected with procurement. The project name "LONG EARS" was assigned. Since the author had presented the opportunity he was designated project officer with wide discretionary powers. It was a heady position for a young Captain.

After appropriate negotiations an order was placed, at a fixed price, with a firm delivery date. Sometime later an extra bonus was offered which caught everyone by surprise. The factory offered to train the buyers' flight crew, at the factory, at no extra cost. Since the crew had not yet been assembled this presented a problem, but one that made everyone very happy. For several years unsuccessful efforts were made to get our intelligence people into Soviet aircraft factories. Now, we were being invited in as their guest. What a bonus!

The requirements for this special crew were very stringent indeed. Finding individuals who could meet these requirements was a formidable task. Needed were pilots, mechanics, aircraft electricians and aircraft instrument technicians who were helicopter qualified and current in their skills. They had to have current security clearance or be qualified to hold one without delay. They also had to be multi-, or at least bi-lingual in the languages needed. Some intelligence collection experience was desired (this was no place for amateurs) but that was not a prerequisite for selection as it might well have eliminated every potential candidate. Finally, they had to be available and willing to accept an assignment that would place them in great danger and could even cost them their lives. Where do you find such men, especially on such short notice?

First the Armed forces were tried. A requirement was levied on the central personnel offices of the Army, Navy, Air Force and Marine Corps to search their records for the names of any individuals meeting this unusual combination of requirements. This was done and, from the entire Department of Defense, a total of eight men qualified for further consideration although none had all of the qualifications. Not an impressive number.

As soon as the name search started, an aircraft with a crew chief, and another pilot, who was a multi-lingual intelligence officer, was assigned to assist. The author is (and was) a bilingual pilot, an aeronautical engineer, a licensed Aircraft and Engine (A&E) mechanic and, of course, an Intelligence Officer. It was a most unusual team, on a most unusual journey, interviewing candidates for a most unusual mission. The team went on standby alert waiting for the first name to fall out of the personnel computers. When the first name dropped, a message was sent from the Pentagon, signed by the Secretary of Defense, directing the candidate's unit to have him, and his immediate superior, available to talk in a secure area immediately upon arrival of our crew at his station. The tail number and anticipated arrival time of the aircraft was given along with the author's name. No other information was provided.

Needless to say, there was no opposition or lack of cooperation on arrival. The reactions were most interesting and, by themselves, make a good story. As an example, one landing was at an Air Force Base serving a large Army installation where an Army candidate was to be interviewed. When the plane taxied in, the ramp crew guided it to the VIP parking spot. The pilots traveled with flight suits over civilian clothes, and upon landing, slipped out of the flight suits, put on civilian coats and ties, and left the aircraft, carrying briefcases. The crew chief stayed with the plane to refuel and prepare for the next flight. All of this, although not done for that purpose, increased the interest, curiosity and tension of those meeting the plane. The airfield operations building was essentially evacuated, with a white rope stretched down the middle of the building and armed Military Policemen stationed every ten feet along the rope. One half of the building had been evacuated for the interview as the "secure area" specified in the Secretary of Defense's message.

A very nervous candidate, a Sergeant, and his Commander, an Army Captain, plus a Chief Warrant Officer were waiting. When asked who he was and what he was doing there the Warrant Officer explained that their Division was currently undergoing their annual Operational Readiness Inspection (ORI) and that a normal part of every such inspection was some unexpected challenge thrown at the Division to see how fast, and how effectively, they responded. The Commanding General suspected that our mission might be that test so he directed the Warrant Officer, from the Division's Intelligence Office, to find out what was going on and report back. We assured him this was not their test and that we were not aware they were having their annual evaluation. He was also pointedly reminded that the message from the Secretary of Defense was very explicit as to who was to be present during our interview. He said he understood and asked if he could attend the interview so he could give a better report to his General. His request was granted with the proviso that, if he interfered in any way, he would immediately be directed to leave the area. He assured us he understood, accepted the conditions, and was permitted to stay.

The interview could not start at once because of the nervousness of the candidate. The Sergeant was afraid he had done something terribly wrong. He didn't know what it was, but knew that it had to be truly bad if a team at this level had been sent all the way from the Secretary of Defense's office to get him. His fears only increased when the first act, inside the interview room, was to set up equipment to record the interview. To put him more at ease, the first fifteen or twenty minutes were spent visiting and getting acquainted. During this time he was assured that he was not in trouble, but was being paid a compliment just to be singled out for the interview.

Then he was told that a team was being assembled for a highly classified, clandestine intelligence operation requiring men of high integrity possessing unusual combinations of very special skills, and that he had been nominated as a possible candidate. He could not be told what the project was, the nature of the project, nor where in the world he would be sent, but it would require extended absence from his family and possibly place him in great personal danger. He was also advised that if at any point he decided he no longer wanted to be considered, he was to say so, the interview would be terminated, and he would return to his unit with no record of the interview. He understood and requested that the interview continue.

The pattern for these interviews included coverage of the candidates' area (s) of technical expertise, his ability in alternate language (s), his emotional stability, his maturity and his overall suitability for this project. The same pattern was used with all candidates interviewed.

Questioning began about his work, his family, his technical specialties and other topics which might bear on the special project for which he was being considered. As the questioning continued languages were switched to see how quickly he responded, and how fluent he was in his alternate language (s). While we talked, the author noticed this candidate was missing a finger from one hand. The finger next to it was oversized to neatly fill the gap so that it was barely noticeable but, in clandestine operations, physical abnormalities can become a major problem. They unmistakably mark a person and can become hazardous to ones life. When asked he said he was born without the finger and its absence had never caused him any difficulty. The candidate and his Commander were then advised as to what would happen if the candidate was selected. He would be sent on temporary duty (TDY) for ninety days, the TDY would be extended twice, then, after he was somewhat forgotten in his unit, he would be quietly transferred, permanently.

When the interview ended all three were again cautioned not to discuss the interview, nor our having been there with anyone except their Commanding General. At this point the Warrant Officer said to me, "Sir, I don't know what this project is, but I would like to volunteer?" Regretfully he had to be told he did not have the prerequisite qualifications.

Following this, a second interview was held with the candidates Commander alone. His candid appraisal of the man's qualifications, his general character and demeanor and, finally, his personal opinion of the suitability of the man for such a job. His Commander gave him an outstanding endorsement.

A call to the office of the Secretary of Defense was made to get the name and location of the next candidate and the interview crew flew directly to that location. There was no time to be lost.

As we were flying to the new location the other pilot, who had not been involved in clandestine operations, said he was caught by surprise as he had not noticed the absence of the finger. When its importance was explained, he understood and was more observant in later interviews.

Note: (After seriously considering this matter, and many others, this sergeant was selected and did an outstanding job).

Only eight candidates were identified by the Department of Defense. One was eliminated but seven were interviewed. The other interviews proceeded essentially as outlined above with one exception. That candidate was crew chief on the Presidents' helicopter. There were no problems with the man, but his Commander (a Major) was very pompous and indignant that his man was even being considered for another assignment since, the mans' current assignment carried the highest possible priority. The Major continued to give trouble throughout the entire process, and arrogantly insisted that his man would not be released, even if he were selected. He was selected, reported three weeks later and became an outstanding member of the team.

As soon as the crew was selected and assembled, aliases were assigned, world-wide backgrounds established, credentials prepared and other parts of a cover story constructed which would fit the established scenario for buying the helicopter. All of this became more important when the invitation to send our crew to the factory for training was received. The crew would be the subject of close scrutiny and every part of the story had to stand up against any background checks that might be made by anyone, including foreign intelligence operatives. The story was so well established and supported that it is still in place and could be used today.

Since the buyers were represented as being involved in business world-wide, the crew had to represent journeymen professionals from the world market. The technical part presented no problem since these carefully chosen men were among the world's best trained and most highly qualified but their background stories had to be established to stand up under any investigation. Accordingly each member of the crew was required to learn as much as they could about the areas matching their story. Following their initial briefing and indoctrination each was sent to the appropriate part of the world to spend time and gain first hand information on their assigned background area.

Another part of the preparatory program was training the crew on the intelligence aspects of their mission. As much as possible, they were prepared for what they could expect, what they were expected to do, and what they should look for while they were on location at the Soviet factory.

On the technical side the best vehicle for training was the US helicopter that most closely resembled the purchased aircraft. There was one aspect they could not be trained for. This was that the rotor blades of the training aircraft and the purchased aircraft turned in opposite directions. This, in turn, meant the controls are reversed. The Army readily provided the training aircraft. The crew had to get as much flight practice as possible and become razor sharp, so the control reversal and other differences would cause minimal problems on site.

It was during this flight training phase of the project that the author was able to enjoy a very special personal experience. The crew was required to maintain the normal flight proficiency requirements of their respective service including cross-country flight. On one occasion the crew needed a cross-country flight and advised their destination and planned flight route and their expected time to be away. The route chosen went close to the farm on which the author's sister, brother-in-law and their family of seven small children lived. His sister had no idea what he did but did know that he flew all over the world, so she was constantly demanding that he "drop in on her" sometime. With this in mind, he asked the crew if they could drop him off on their way out and pick him up on the way back. They asked if there was enough room to set the helicopter down. When informed that there was 150 acres of farm backed up by a State Forest, they said, "Why not?" Since our parents lived about twenty miles away from the farm they were called and asked them to be at the sister's house at the estimated time of arrival (ETA). When they asked why, they were told not to ask questions, just be there. The sister was not called.

As the helicopter flew into the area the author took the controls and located the farm. Then the regular crew took over, picked out a landing spot and sat the craft down. As it settled to the ground, in the barnyard, the mother, father, sister and the seven kids lined up at the fence, with the kid's eyes looking like saucers. (Really, the adults' eyes weren't much smaller???). As soon as the author stepped out, the helicopter lifted off and went its way. After the excitement died down, there was a nice visit and, four hours later, the helicopter returned, and we returned home. Later, the sister said calls were received from neighbors, over a period of four days, wanting to know if that helicopter had crashed. Never, to this day, has she repeated her demand that her brother "drop in on her". It remains a happy memory of a rare fringe benefit.

A very critical requirement when documenting the crew was their flight credentials. Fortunately, about the only flight certificates unquestionably accepted throughout the world, at that time, were those issued by the US Civil Aeronautics Authority (CAA), which had certifying offices in various locations around the world. Through extraordinary cooperation from key individuals in CAA, certificates were issued for the crew, reflecting the appropriate field offices, with the appropriate names and backed up with the proper credentials, paperwork and signatures so that, even today, they would withstand any background check. These documents established the crew for what they purported to be, and were accepted without question. The author, personally, spent several days working with CAA officials preparing these documents and cannot praise them enough for their complete, unreserved, and willing cooperation. Without them, the job would have been much harder, or maybe even impossible.

The US Department of State also provided everything asked for, including the necessary passports and documents.

Midway through the process of selection, training and documenting the crew a very unexpected message was received from the factory-the helicopter was going to be ready for delivery three months ahead of schedule! The invitation to send our crew to the factory for training was repeated, only three months earlier than planned. Unfortunately the crew could not be ready to go on the accelerated schedule, so the factory was notified that the new schedule could not be met. They in turn, trying to satisfy their new customer, offered to send the helicopter and factory crew to a third country of our choosing where delivery of the aircraft and crew training could take place, at their expense, thus meeting our need for a delay. There was no plausible way to decline their very generous offer and still maintain our position as a commercial buyer, so the offer was accepted and new plans very quickly made.

The helicopter and both crews arrived at the delivery site on time. The crews worked together to assemble the aircraft, test fly it and train our crew. With two minor exceptions our crew was instructed to go about their business in the normal way speaking the languages they knew. One man, who spoke fluent Russian, was instructed not to let the Russians know he spoke their language but listen in on their conversations and report back to us. This worked as the Russians relaxed and spoke freely when he was the only one around. What he heard helped us avoid trouble spots, and identify possible security weak spots.

Another of our crew, a mechanic, was an older man who looked like the stereotypical US movie version of a German. He spoke the language fluently in a guttural voice but told no one that he spoke English. He played his part to perfection, always grousing about everything, in German, as he went about doing his job very well. Some of the Russian crew members spoke English and would make derisive comments about him in both Russian and English. He became the butt of all the jokes, and a relief valve for both crews. It made everything easier, and the crews got along very well together.

One thing that surprised us was the total lack of training and skill of the Russian crew in handling sling loads under the helicopter. This is one of the most difficult maneuvers for a helicopter crew and the Russians had received very little or no training in this area whereas our crew was highly skilled.

The entire operation of assembly, flight testing, crew checkout and training was completed without incident. Then the helicopter was disassembled and recrated, including the spare parts, for shipment to the designated destination. The crews shook hands, said their farewells, and went their separate ways.

The crates were assigned for shipment to a fourth country, but were routed for transshipment through a major US Port. At every major international land, sea or air port there is an area called a "Free Port" where a shipment can be off-loaded for transshipment on another vessel without legally entering the host country. These shipments are not assessed customs duties, or taxes, since they do not legally enter the country. Some tradesmen, who buy foreign goods for sale in the country, will pay storage in the free trade zone for their surplus stock, then pay duty and bring it through customs as they need it and after they know the item will sell. All countries have such areas. Security in a Free Port Zone is very tight to prevent smuggling of the goods into the country without paying duty.

There was no problem with paying duty, but it was our desire that the contents of our shipment "disappear" from the free port while the crates continued on to another destination and nothing be on record as having entered our country. The Chief Officer of the port was briefed on what was being done and what was needed, then asked for his cooperation. After some discussion he called the chief of his anti smuggling division, told him what had been discussed, then directed him to help. The division chief was disturbed for he said, "I have one more year until I retire and now you are asking me to do something that I have devoted twenty seven years to prevent happening." But, to his credit, he pitched in, the crates were emptied, reloaded with ballast of about the same weight then put on another carrier going to the destination shown on the manifest. Meanwhile, the original contents were recrated and shipped to a remote location within the US for exhaustive testing and evaluation. Thanks to the complete cooperation of the port officials this all happened without a ripple. How can you ever really repay people like this?

While this was happening, the crew was given vacation time off and told to prepare for transfer, with their families, to the remote location where they would check out our best test pilots and assist while a complete flight test program was conducted on the helicopter.

At the same time, in a most unusual move, the author was transferred to the same test location. Among other duties, was monitoring this test and evaluation program for the helicopter. This was very unusual indeed. The author proposed acquisition of the equipment, planned the entire program, set up the acquisition mechanism, directed procurement, selected the personnel involved, was overseer during the factory training program, coordinated the shipment and now was being assigned to monitor its test and evaluation. He had overseen the project from its conception until it's completion. Who could ask for more!

Much preparation and planning is necessary to maximize the gain from a project of this magnitude. One area that could have given many headaches was working for so many bosses. It seemed that every branch of the government had a vested interest in this project and its' findings. Many organizations were putting money and other resources into the project and they wanted the maximum out of it. Fortunately problems in this area did not materialize. All concerned readily provided the help needed, stayed out of the day to day operation and allowed the project to go forward at optimum speed.

Preparing the test plan was an excellent example of the rapport among all concerned. The author chaired the planning sessions which included carefully chosen representatives of the Army, the Navy, the Air Force, CIA, DDR&E (Deputy Secretary of Defense for Research and Engineering), often the Marine Corps and others whose interests were strong enough to be included.

Always included, of course, was the Air Force officer who was the on-site, hands on, test director. No better man could have been selected. He was a test pilot and a flight test engineer who had been a U-2 pilot (circa Gary Powers). He had all the skills, attributes and characteristics that were needed to do a near-perfect job.

Even on this usually very serious and tedious project there were some interesting moments. The test site was near a National highway and there was concern about unwanted sightings during test flights. At one meeting this problem was being discussed. CIA was very helpful throughout this entire project and had two representatives, a mature technical man and a young, fairly new, man apparently being groomed for bigger things. As various ways of meeting this potential problem were discussed the younger CIA man spoke up and said the solution was simple, just let them know when the plane was going to fly in that area and they would plant two explosive charges, blowing up the National highway on each side of the test area. There was stunned silence among the other twenty five people present before the meeting proceeded to other things. Flight operations were conducted without any problems but can you imagine the media attention that would have been drawn to the test area if there had been explosions blowing up two sections of a National highway? The young man was relieved of his duties and was not seen again. Such things kept life from becoming dull.

Assigned to the project was the best helicopter test pilot, anywhere, at that time (He held several international records.) and the test results showed it. One day the author was riding with him, on a routine test flight, when he said, "Here, you take it", and there he was, in "hog heaven" flying this very special bird. Surprisingly the reversed controls gave no trouble. This was a special treat even though the test pilot was ready to come to the rescue if something did go wrong. The bird had to be preserved at all costs.

The test and evaluation program proceeded smoothly because of the quality of the people assigned. They were the best, including the hand-picked guards assigned to provide security. Even so, there were problems. One morning, at 2 AM, a call was received advising that someone had tried to break into our remote facility, and the guard had fired through the door to chase the intruders. There were twenty project people on location at that time. All were called out to assess what, if any, damage had been done to the equipment or the project. So, twenty very highly qualified specialists, along with an elite unit of guards, were scouring this remote area in the wee hours of the morning trying to determine what had happened. Project people were kept on hand until mid morning before finally being released to go home and get some rest while the professional investigators continued their work. That work day was lost.

The next afternoon the guard commander called and asked if we could talk. He advised that intensive interrogation of the guard revealed that the young guard on post had become bored and started practicing "quick draw". In the process, he had accidentally fired off one round which, luckily, had gone through the door. Since the guard had to account for every round of ammunition, he concocted the intruder story to cover himself. The guard was immediately suspended and shipped off to another assignment where he could do no harm.

There was another incident. The test area was blocked out as a "no flight" zone, which meant that there was to be no overflight of that area by anyone. The local Commanding General was briefed regularly on progress and problems encountered. At one of these sessions the General asked if there had any trouble with overflying. He was informed that there had been one incident. He immediately wanted to know the name of the individual. He was told, "General, it was you". He was startled but wanted the details of the incident, which were provided. After thinking a few moments, he said, "By golly, I did, didn't I"? He then asked how we handled it. He was told that, when he landed, a member of the project collared and debriefed every one on his plane, except him. He complimented us on the way it had been handled, and assured that he would never again violate the zone, but that if he or anyone else did he wanted to be advised immediately. The General was well liked and respected.

As the project proceeded the gains were identified and began to flow out to the users. One of the first identified was the rotor blade de-icer system which proved to be much more efficient and more effective than our best. This was immediately passed on to US designers.

Next was a study of their paint. Earlier observations of Soviet operations indicated that the Soviets had far less loss and damage due to corrosion than the US. We wanted to know why. Since we had acquired many gallons of paint in the fifty per cent spares we could, and did, commission extensive salt water spray tests of the Soviet primer coat alone; the Soviet primer plus finish coats vs the best US primer alone and the best US primer plus finish coat. These tests didn't last very long for they quickly showed that the Soviet primer alone still showed no signs of deterioration or corrosion after it had tripled the breakdown time of our best primer plus finish coat. Simultaneously commissioned was chemical analysis of their primer and finish paints. Thirty one different elements were identified along with the percentage of each in the total paint composition. This allowed almost immediate duplication of the paint for use by our forces. The Navy advised that this item alone would save over $6,500,000 per year (in that year's dollars) in repainting costs for their ships.

Another extremely valuable gain was the ceramic coating used on the exhaust stacks of the engine. It was known that the infrared (IR) signatures from all their aircraft were very low. Initial studies verified that the IR readings from this helicopter engine were substantially below those of our nearest equivalent. The importance of this information was that many of our best anti aircraft weapons relied on Infrared (IR) guidance systems. The weaker the IR signature, the less effective does the guidance system work. So, learning the reason for this low signature was very important as it could apply to many other weapons systems and our ability to defend against them. Extensive tests traced the low IR readings to the ceramic coating used on the exhaust stacks (pipes) and other very hot parts of the power system. Again, the spare parts included buckets of this ceramic coating which was immediately analyzed, and duplicated for use by US forces. A huge gain.

There were many other items which proved of great operational value and/or great gain to the USA. As each item of potential gain was passed on to the ultimate user. They were asked to assess the dollar gain that the information gave and pass the information back to the project office. When the identifiable gain reached $30,000,000 Gen Watson directed that the adding should stop. He very correctly stated that reporting a "$30,000,000 gain from a $1,800,000 initial investment" would be very hard for the powers that be to swallow. (Any more and no one would believe us). What a happy position to be in.

As the project came to its conclusion the Major General (two stars) commanding the test location was asked to prepare and present briefings at the highest command levels of the Pentagon and other National locations. He directed that a briefing be prepared and given to him in rough draft form so he could put it into his own words. He then added that he thought that the author (still a Captain-two bars), not him, should be giving the briefing since the author knew far more about the project than he did, but then he added that others thought "two stars carried more weight than two bars" in the Headquarters arena. No one could disagree with that. The author was to accompany him on the briefing trip, run the visual aids, then answer the questions for he said he knew that he would not be able to answer all the questions asked. Several very successful briefings, one with over thirty Generals and Admirals present, were given and many accolades received. A heady place for a young Captain. This General, now deceased, was a favorite.

As this excellent project came to a close, the author thought the outstanding performances by members of the team deserved to be recognized in tangible ways. To accomplish this commendations and recommendations on the five men (four Army and one Air Force) who constituted the original crew and had stayed with the project throughout it's life, were forwarded through channels for approval. The recommendations were accepted and acted upon. Each of the five received Presidential-directed, noncompetitive, promotions and each received an appropriate decoration.

Because of the intense security surrounding this operation the medals were presented, by the Air Force General commanding the test site, in a private ceremony in his office, with only the men's wives and project personnel present. Although this was a solemn occasion there was one very humorous incident. All parties present for the ceremony were waiting, a bit on edge, in the authors office until time to move to the General's office for the presentations. At that time there was a young lady working in the building, dearly loved by all, who was a comedienne in her own right. She decided to drop by the office for a casual visit and, in her usual style, swung the door open and danced into the room singing "Ta-Ta", then suddenly realized that about twenty people, dressed for the ceremony, were setting there solemnly looking at her. You've never seen anyone so embarrassed. She backed out of the door stuttering and blushing. As she left, the room exploded with laughter, the tension eased and everyone relaxed. Later the honors were presented in a quiet ceremony without further fanfare.

Not long afterwards, when this young lady married the General's driver, the staff gave the young couple a wedding gift consisting of several cases of canned goods, all with the labels carefully removed, to help them start their new home. Great fun enjoyed by all and, as you can see, it was not all sober and dull.

Unfortunately, the Air Force member of the original team died a few months later.

A short time later Army representatives asked, "What kind of people do you work for"? When asked the reason for their question they replied that the Army had recommended to the Air Force that the author be appropriately recognized for his role in conceiving this project, planning and organizing it, directing the acquisition and finally the exploitation to such an imminently successful conclusion. The Air Force's answer was "Why, he was just doing his job"?????

Such is the fate of an intelligence officer. Recognition comes only when a mistake is made. Recognition, such as the Army recommended, would have been welcome, and gratefully received, but the greatest reward is knowing that his country was served to the best of his ability, there were no casualties, all had gone well and much value had been gained.

THIS ONE WAS A WINNER.

MI-6

Information on new developments and technology was constantly being gathered from many sources then analyzed and evaluated to insure that potential antagonists would not gain either technological or operational advantage over the United States. This process included, but was not limited to, identifying equipment that could be valid targets for acquisition and exploitation through the Foreign Material Program. A prime example was a very large helicopter, estimated to be capable of carrying thirty to fifty troops, which was the subject of many intelligence reports but had not yet been sighted. The reports came from areas of known Soviet experimental aircraft activity, and were submitted by reliable intelligence sources and were taken seriously. However, when leading US helicopter manufacturers were consulted the reports were ridiculed as being preposterous. The reason given was that the rotor blades necessary to support such a craft would have to be twice as long as the longest US blade in use at that time. It was further emphasized that the US blades were the longest that could possibly be made to effectively fly on a helicopter. This was a classic example of the attitude often encountered in our scientific, engineering and manufacturing communities during the early days of Scientific & Technical Intelligence. "If we (the USA) haven't done it, it can't be done" was the prevailing attitude. Of course we also wanted to believe this but, too often, we had to first prove our own people wrong before we could get their cooperation. In later years, after we had proved ourselves, the opposition essentially disappeared and all parties accepted the fact that we had no monopoly on brains, skill and capabilities.

It was in this time frame that Premier Khrushchev of the USSR, during a visit to the US, rode with President Eisenhower in one of the President's new Sikorsky helicopters which was equipped with our newest development, automatic-pilot flight stabilization equipment which provided a much smoother ride and also made possible instrument flying during bad weather. Without such equipment night and instrument flying is very difficult in helicopters as there are no visual points of reference to help maintain straight and level flight. The Soviet Union had been unsuccessful in their efforts to develop such equipment. Premier Khrushchev greatly admired and praised the helicopter and asked the President if it would be possible for him to obtain two for the Premier's use.

This caused much concern because the new equipment on helicopters gave the United States a very distinct advantage over their Soviet counterpart. Much effort had been expended to keep this, and other items of advanced technology, out of Soviet hands. However another, overriding, factor entered the picture. In Diplomatic circles such a request, from one Head of State to another is almost never turned down, as it might cause international tension, so President Eisenhower felt obliged to grant Premier Khrushchev's request. The Soviet

Foreign Materiel Program was obviously working very well as they were recruiting their collectors from the very highest level—the Premiers office. Aware of the reported (but not yet seen) extremely large Soviet helicopter a suggestion was initiated (while there was still time to do so) that President Eisenhower make a counter proposal in which he would praise the Soviet development of such a large helicopter and suggest that there be a trade the Sikorsky (s) for one or more of their big Mikoyan helicopter which had by then been assigned the NATO designation, MI-6. We would still have to surrender our protected technology, but would gain something of value in return in the form of their ability to make extremely large blades work efficiently which we had been unable to do.

Unfortunately that recommendation apparently never reached those in a position to take the necessary action. The author received no response and, some months later the two Sikorsky helicopters were delivered to the Soviet Union. The United States surrendered our technology and received nothing in return.

Later, after President Kennedy had been elected but not yet taken office, on a visit To the Headquarters of a major Air Force installation. the author met Gen Watson (the founding father of S&T intelligence) entering the building accompanied by a very sharp looking young man. Greetings were exchanged then the General asked that we go back into the building as he had some things he wanted to discuss. After making our way through two vault doors into a secure area the General introduced his companion as the man who would be President Kennedy's personal White House intelligence advisor when the new administration took office. He then told the new advisor who I was, what I had done and was doing, in terms which increased the hat size at least six numbers. A number of projects, current and past, were discussed with great interest expressed by the new advisor-to-be. The new man then asked if any occasion had ever arisen where quick action was needed and could not be obtained. He was then told of the helicopter swap recommended which was neither acknowledged nor acted upon. Gen Watson was very disturbed and wanted to know how the proposal was submitted. He was reminded that it had to be up through channels. He responded that it never reached his desk, that the recommendation made sense to him and, had he received it, he would have acted on it. Knowing him, I am certain he would have.

Hearing this the young man wrote a telephone number on a piece of paper and handed it over with the comment "This will be my personal number in the White House. If anything like this comes up again, don't worry about channels, call me direct. We cannot risk passing up opportunities such as this". An occasion to use the number never came and we also never managed to get one of the large troop carrying helicopters in our hands for study and evaluation.

SECRET SHADOWS of YESTERDAY

PROJECT BIG TOM

Another opportunity to get one of the very large Soviet Mi-6 troop carrying helicopters did not materialize. However, infromation began coming in about a "flying crane" version of the MI-6 which was essentially the same aircraft but without the very heavy troop-carrying passenger compartment. Known as the MI-10 this was of immediate interest as it was at least twice as big as our biggest helicopter crane and potentially offered twice as much heavy lift capacity. Having such capability could provide a distinct advantage. Interest in this big bird was very great and an immediate effort was launched to get one.

Operating and testing such a big helicopter, if one was obtained, would present and entirely new set of problems. The big blades would make such a powerful "whomp-whomp-whomp" sound that it would be nearly impossible to operate and test it anywhere without attracting attention. The noise level of engines powerful enough to turn such blades would only increase this problem. So a new program had to be devised which would allow procurement of the aircraft, completion of the test and evaluation program, and eventual disposition of the remains without drawing unwanted attention. How to do it?

Then the obvious answer hit. The size of this aircraft with its' potentially very heavy lifting capability should make this helicopter crane of great interest to any industry where heavy lift capability is both needed and wanted so such a company was sought that would cooperate in procuring one and operate it at their work location (s) with "special interests" participating behind the scenes.

An immediate search was initiated to find such a company. Several were identified and their day to day operations, their personnel and their operating location (s) studied to find one suited to the needs, wants and desires. Again a pleasant surprise as more than one met the requirements. After careful study one was selected and approached about purchasing this huge helicopter in their name, having it delivered to an agreed operating location, operate it in their normal business with some special conditions applying. These conditions included providing the purchase cost, the operating costs and the cost of hiring "special" employees to work within and help run their normal operations plus some that the company might not otherwise perform. These "special" employees would, of course, be our very highly qualified test people.

No operation of this scope could be done "under wraps", so many hours were spent trying to identify and plan for every eventuality. Among these was the possible objection to buying "Russian" equipment instead of "buy American". Press releases were prepared to cover every conceivable situation.

A decision to inform the US Representative in that district and ask for his cooperation and support was made. This was done somewhat reluctantly as he had the reputation of not being too friendly to, or supportive of, the military and its' operations but it was also thought that political support of such horsepower might be needed. He was briefed on the planned program, the expected gain to the United States and the need for absolute security concerning any US government connection with the company's operation. It was also pointed out that it would pump a lot of money and employment into his district. Possible needs for his participation were discussed and how he should react if certain undesirable events occurred. He was intrigued with the project and happy to have the project located in his district, cooperating in every way and presenting no problems. Press releases were prepared, with his cooperation, to cover any occasion. This Representative, now deceased, proved to be a pleasant surprise.

The Project proceeded very smoothly and maximum information was gained in minimum time. It helped to know that such a high priority major undertaking coul be conducted wide open, without security breaches at a most economical cost figure. A very valuable lesson indeed. It was also confirmed that the very large blades which "couldn't possibly fly" flew very well.

While visiting the site and personally examining this huge aircraft it seemed when sitting in the cockpit, over thirty feet in the air felt as though the aircraft was at cruising altitude although it was still setting on the ground.

A major concern during original planning was how to dispose of this huge, one of a kind, aircraft when the project was finished. A plan was agreed upon and a press release written that would cover the situation and, hopefully, keep questions and inquiries to an absolute minimum. Interestingly enough, five years after leaving the project that press release appeared in a newspaper in another part of the country.

All's well that ends well.

FIGHTER

All stations were always alert for new opportunities to get the latest technology of our opponents. This paid off when it was learned the Soviet Union was planning to equip a favorite Client State with one of their latest Jet fighters. When this was learned an immediate major effort was mounted to get one of the aircraft.

Several possible plans were considered. The one selected had the potential of rivaling some of the best spy stories you might read in pulp fiction. The plan was to recruit one of the Client States' pilots to defect to a nearby country while flying an assigned mission. But, before recruiting the pilot, many other plans and arrangements had to be made and implemented.

First need was a local person or group in the neighboring country to manage the in-country part of the operation. They would be responsible for finding an abandoned airfield equipped with a suitable hanger which was within the aircraft's flight range. Next they would have to assemble the necessary technical personnel, such as aircraft mechanics, to properly service then disassemble and crate the plane for shipment. Also needed were shipping arrangements that would not raise any unwanted attention. Finally, and most important was very tight security arrangements for the whole operation. They would, of course, be very well paid for these difficult tasks.

The master plan called for the pilot to defect, preferably near dusk, by flying at a very low altitude to the designated airfield, land, and taxi straight into the hanger which would have its' door open with no lights on. Once inside the pilot would shut down the engine and the hanger doors would be closed. The author was to be on location to make positive identification that the plane was the one wanted. After confirmation of the aircraft type the pilot would be escorted from the premises and paid. Next the mechanics were to be instructed on the "break points" of the aircraft which would allow easy and proper disassembly and crating without damage. Only after the author departed the area and arrived safely at a neutral international center would the money be released to pay the various individuals and groups involved. This insured personal safety all around since no one, except the pilot, would be paid until all were safe.

In the meantime after disassembly the plane would be crated for ocean shipment, marked "Machine Parts" and shipped to a designated port. There the crates would be loaded on a chartered vessel which would then sail, nonstop, to the desired destination. The US Navy would track the ship to its destination to insure against unwanted off-loading of the "Machine Parts".

This backup required for a defection is very complicated and must work perfectly the first time. There is no margin for error. Appropriate, trustworthy, "in country" people must be found to head that part of the program; they must be selected, informed, motivated and cooperative. The tasks assigned these people take a lot of time and effort and must be done without compromising any part of the operation.

After the aircraft type was confirmed, the pilot was to be paid off and told he was on his own. The equipment collection plan did not include the capability of handling personnel defections so this had to be kept separate from the equipment, which was the primary target. At the same time another agency would be waiting for him as he walked around the corner and separated from the equipment. Information from him could be gained through that agency at a later date.

Next came one of the biggest jobs, gaining final approval and the high six figure financing necessary to do the job. Interest level was so high the money was allocated almost immediately. Concern was expressed about the personal safety of the author under the plan. The main concern was going into the "Lions Den" to deal directly with the Sources and Middle Men when confirming the aircraft type. Although such exposure is not often risked it was believed enough safety measures were in place to insure safe return. There was no doubt in the authors mind that it could be carried off and he was ready and willing to do so. The main safety valve was that no one involved, other than the pilot, would receive any money until he came out of the operational zone to the selected International Banking Center and released the balance of the funds. Since their incentive was money these people would make certain that no one harmed him before they were paid.

The desired pilot was quickly found and recruited. He was given no details except that he was to take the plane and fly it out of the country for a very generous sum of money. He was put on standby while the rest of the plan was being perfected.

In the country where the airplane was to be delivered some political, military and port authorities were "persuaded" to cooperate with us for handsome retainers. They had to be very carefully selected for their part was key to the whole project.

All planning was completed and several parts of the operation initiated such as recruiting the pilot and having the money in the bank but we still did not have the final approval to go. Much vacillating, hemming and hawing was detected and, as it turned out, several cases of cold feet. There were still some strong reservations about the potential risk to personal. The real obstacle finally revealed itself. One General, newly appointed to his position, wanted another star so bad he could taste it. In reviewing the proposed operation he suddenly "discovered" that this operation was in the "Black Area" of intelligence when the

Air Force had no Charter to be in that area. He was technically right but there had been operations in that area for several years because the organization that did have the Charter was not able to satisfy the needs. The General was afraid that if anyone challenged our operating in this area he would not get his coveted star. The organization that had the Charter was well aware of what we were doing and were actually working with us on the project. They presented no problem.

The final result was this General canceled the project and withdrew the money. His action cost us a golden opportunity and almost two years of sweat and toil. The only consolation was, the General never got his coveted star. There was some justice after all.

MEDIUM TRANSPORT

Intelligence collection has a lot of value and is used for purposes other than just getting information about an opponents' particular system or their capabilities in given areas. Occasions do arise where collection efforts are directed towards improving our overall collection capability.

Another Agency asked for assistance in procuring a medium sized Soviet transport aircraft to be used in an operation. This aircraft was a fixed wing transport hauling about twenty five or thirty people. It was heavy, by US standards, as it was built to operate from unprepared fields. For this reason it had a much higher basic weight than our equivalent but, in turn, it could operate effectively out of fields where our equipment could not. Maintenance costs also might be lower because of its construction but it would not be an economical airplane for commercial operations.

This airplane was of no interest for it's technical value but we were happy to assist our sister Agency by acquiring one. It also gave another chance to exercise our procurement channels. Purchase arrangements were made to buy the aircraft, new from the factory, with delivery and transfers through other countries in such manner that its final destination and purchaser could not be identified.After several transfers the buyer took possession at a foreign location. It was suggested that two of the World's finest transport test pilots (who were on standby for just such a mission) take possession and fly it to the buyer's final location. This was logical as these men were among the sharpest pilots in the world and this would have been a routine flight for them since, as a part of their normal duties, they flew new and unfamiliar aircraft almost every day. It offered an additional benefit because they would be available to check out the buyers' pilots in a more routine manner.

The buyer declined the offer. They were very anxious to get their hands on their new purchase. A bit of "Professional Pride" was also suspected but their decision almost led to a catastrophe.

The destination was a major US air terminal where the buyer had their own facilities. Here the flight crews and maintenance people would have the opportunity to become acquainted with their new possession and the plane would be repainted before going on to the planned clandestine operating location. To reduce exposure and unwanted attention the plane was to land at 2AM when activity at the airport would be at its lowest ebb. The hanger doors would be open with the lights off, and they would taxi the plane inside before shutting down the engines. The hanger doors would then be closed before the lights were turned on. Altogether, a good plan.

All was going well until the plane came in for a landing. The pilot was either "flying scared" or up tight for he came in much too fast and, in order to stop on

42

the runway, rode the breaks hard, blowing out both main landing gear tires. So this special plane which didn't want to attract attention was now down in the middle of the runway of a major airport. Fortunately the ground crew was most efficient and got it under cover before any more damage could be done.

The operation survived but two of the four spare tires that came with the new aircraft had to be used. This incident was so unnecessary and it is not likely it would have happened had the recommended crew been used. Mark up another eventful but successful operation.

VISITING TRANSPORT

Mr Krushchev, at the time he was Premier of the Soviet Union, was perceived to be a very crude, unsophisticated, and unpolished, man. This image, based on his pounding his shoe on the podium during a session of the United Nations, was enhanced by reports received from a fellow intelligence agent, a native-born Russian, who once acted as interpreter for President Eisenhower and Premier Krushchev. The agent said the Premier's normal conversational speech was so vulgar, crude and profane that there was no way he could give a literal translation in polite, mixed company. It was so bad that doing so might have precipitated a diplomatic crises. Therefore it was necessary to listen to what the Premier said, put it into clean, polite language then word the translation without delay. No doubt it was a most difficult translating job.

In writing about the possibility of exchanging two Sikorsky helicopters for the huge Mikoyan MI-6 helicopter, in an earlier chapter it was questioned whether Premier Krushchev, because of his lack of the diplomatic niceties, would respond in a normal diplomatic way to the proposal even if it were made by the President. Opinions were divided.

However, on at least one occasion Premier Krushchev showed that he could when he sent President Eisenhower a military cargo plane full of gifts from the Russian people. When flights such as this arrive in our country, it is customary for high officials to meet and formally welcome the accompanying officials and accept the gifts being offered. Because of the distances and time differences involved standard practice calls for the plane to enter our country the day before, land at an an outlying point, and remain overnight to give the visiting diplomats, and their crew, the opportunity to rest. This also insures the plane arriving in the capitol at the scheduled time carrying well rested officials. This procedure was followed for this fight.

The first incoming flight advisory received through air traffic control informed they were sending their newest military transport aircraft and it was to land, and remain overnight, at an international airport located in one of our major US cities. Prior to this time access to this aircraft, which was of great interest, had not been possible.

When the first information was received the author, along with another trained field intelligence officer, were flying across the United States on another mission. A message was relayed diverting us to the named city, to arrive after dark, remain over night and follow further instructions which would be transmitted to us, by secure means, after we were on location. This presented problems. Expecting to be home that night we didn't have overnight kits nor any clothing except our military flight suits. Adding to that the weather was very bad-a cold drizzle was turning into freezing rain. Whatever the mission, we were in for a miserable night.

On approach to the airport the weather closed in and we were flying blind in heavy clouds, freezing rain and snow. All seemed well as we were under positive ground radar control. Then, without warning, the ground radar failed and we were thrown back into the old system of being directed to "holding" points over ground radio navigational stations. Using this equipment, radio direction finding equipment in the plane homes in on the selected radio stations. This was going well enough until our planes' direction-finding radio also failed, leaving us flying blind, with neither the ground control people nor the plane's crew knowing our exact position or proximity to other aircraft in the area. Not a comfortable situation. At last we were able to get our aircraft radio working again and made our way to a safe landing.

On arrival we were told about this most interesting new plane. The large four engined turboprop, had already arrived and was pointed out to us before we left the airport. Our mission was to get all the information possible about this airplane from the outside. The co-pilot and I were both aeronautical engineers and experienced field collectors so knew exactly what had to be done. The work would be difficult for we had to do our job then be off the ground, and out of sight before daylight since we were flying an airplane carrying distinctive markings which, if seen, might alert the Russians as to our presence and mission.

The guest aircraft was under constant surveillance and it was recommended by the local crew that we not start our work until midnight. By this time, hopefully, the visitors would be bedded down for the night. Our crew had already put in a long day so our hosts had thoughtfully reserved rooms in a local hotel to give us a chance to have a good dinner and get some rest before midnight. However the hotel dining room required coat and tie to get service and, since we had only our flight clothes, the Maitre de would not allow us to enter. The hotel did not have room service but finally we were able to get something to eat. A very long day.

At midnight the local crew picked us up and proceeded to the airport. En route we were advised of still another problem. The Soviets had left one man on board the aircraft for the night. This greatly complicated the mission since we planned to crawl over the plane checking on many different items and taking many measurements such as the wingspan; length; height; the diameter of the engine tailpipes; as well as the length, breadth and curvature of the propeller blades among other things. All this had to be done while a guard was sleeping inside. Hopefully we could complete our mission without alerting the occupant.

The local crew provided ladders, measuring tapes and other equipment needed, then pitched in to help. Their cooperation was most welcome. The night was very dark, cold, wet and miserable. Since there was no heat in the plane, the guard was probably heavily wrapped to keep warm and possibly he was also well supplied with vodka. Whatever the reasons, he never showed himself nor indicated he knew anything was going on even though we had one man walk the full length of the wing and had others on ladders to each of the four engines and other parts of the airplane. All information desired from the outside of the plane was obtained, the operation ended and departure from the airport accomplished before daylight without incident. This operation provided the first solid data we had on this new military transport and the information gained satisfied our immediate needs.

The next day the transport arrived on schedule in the nations' capitol, the guests were welcomed and the gifts received. All went well.

GROUND TO AIR MISSILE

It is acknowledged that the United States is the world leader in so many scientific and technical fields and in our research and development skills. It is not surprising that our opponents have great interest in our new developments and have maintained massive collection organizations targeted on the USA.

We never underestimated the capability of those collection organizations therefore it was not a surprise when we became aware of one Soviet Bloc success story. We learned there was a complete, operational US Air-to-Air missile setting in Prague, Czechoslovakia. Trace action could find no missing missile of this type in either the production cycle or in the military inventory. But a complete, operational missile was setting in Prague. Where did it come from and how did it get to Prague???

We never found out how they did it but we had to acknowledge their proficiency. It was a daily battle on both sides.

FLIGHT OPERATIONS

While serving as pilot for the High Commissioner there were other flying duties more closely aligned with the primary reason for being in Austria. They were also more closely aligned with the things from which spy novels are written, and as you will see the truth is sometimes better and more interesting than fiction.

During the occupation years, Austria was partitioned into four Zones. Each Zone was ruled by one of the occupying powers: the United States, the USSR, France and Great Britain. Vienna, the country's capitol, was situated in the middle of the Soviet Zone about one hundred miles from the American Zone and the city was divided into five Sectors. Each Sector was under the control of one of the four powers with the fifth one being the International Sector containing the Central Governmental offices. The arrangement was very similar to that of Berlin, Germany but differed significantly in other respects. Control over the International Sector, and defacto control over the entire country of Austria, changed hands each month. This unique arrangement produced some very unusual ceremonies. On the first day of each month a formal changing of command was held in the International Sector with each of the participants, except the US, trying to outdo the other for Pomp and Ceremony. The Soviet Union would usually fly in the Red Army band from Moscow; the French would often bring in their Alpine Division unit in which all band members played the French horn; and the British would bring in their most colorful units, such as the Scottish Highlanders all bag pipe band. For its part, the US usually used the US Army unit band stationed in Salzburg, Austria. They were very good but not as spectacular as the others. These events provided some memorable occasions.

A very unusual feature of this division of powers and cooperative rule was the policing of the city. Every police function was a five nation operation with the nation in control of the International Zone being the host for that month. This was most evident when police patrolled the area. The cars always carried five policemen: An American, a Russian, a Frenchman, an Englishman and an Austrian. You couldn't get away with anything in that city. This arrangement was so unique and so interesting that an American movie was made about it.

An even greater difference between the two cities was the constant head-to-head confrontation and publicity in Berlin. Also the demarcation lines between the occupying powers' sectors were sharply defined with travel between them rigidly controlled. By contrast Vienna was so relaxed that only the people living there knew where the Sector dividing lines were and there were no restrictions on travel between Sectors. There was one theory that the real reason for this difference was that both sides in the Cold War ran extensive intelligence operations through Vienna, therefore there was something like of "Gentlemens

Agreement" to have as little confrontation as possible and to look the other way when something a bit unusual was happening. It is only opinion, but it would explain some of the things that happened (or didn't happen). It certainly made operating there much easier.

The USA and the British each had, and controlled, an airport in the Soviet Zone near Vienna and the Russians had two. The French did not have one. Providing access between Vienna and the other Zones there were air corridors ten mile wide in each occupation Zone where the other occupying powers could travel by ground or air unchallenged. The US had a very nice Douglas C-47 (Gooney Bird) for the Ambassador and assorted light aircraft, such as the L-5, L-19, L-20, L-24 or whatever else was available for local flying. These aircraft were used for many things including some of the most hair-raising experiences of the author's life.

THE NICE AND PLEASANT

Flying as an intelligence officer could, and did, take many forms. Some flights were very nice and pleasant; some were routine or mundane in nature; many were very interesting; and others were dangerous, even hazardous. In the course of a career all of these were encountered but, since the love of flying was so strong disillusionment, discouragement nor boredom ever became a factor. Every minute of every flight was enjoyed.

One of the most pleasant and, at the same time most interesting, flying assignments a pilot could have had was serving as personal pilot for Lewellyn Thompson, High Commissioner to Austria during the 1950's. (The High Commissioner carried the rank of Ambassador but a different title since, due to the four power occupation of Austria, the United States did not at that time have formal Diplomatic relations with Austria. Ambassador Thompson was the US diplomat who negotiated the first two successful negotiations between the Soviet Union and the United States following World War II. First the Trieste settlement and later the Austrian peace treaty brought freedom to the people. This assignment was especially enjoyed because it brought direct contact with Ambassador Thompson, a man greatly admired and respected. No matter what happened he never became ruffled or disturbed and he was always most considerate of those assigned to work with and assist him. He was a true Diplomat in every respect.

One example. The Ambassador was returning to Vienna by way of Paris, France after having been called to Washington, DC to meet with the President. Since his home was behind the Iron Curtain we were scheduled to pick him up in Paris and return him to Vienna. The plan called for the plane to arrive in Paris the afternoon before and be prepared to depart at a set hour on the scheduled morning. Arriving as planned without incident the crew spent the night. The next morning Three things governed all activities. First was the time set for departure for, with a passenger of that rank and status, you are not late. Second, at that time it took at least one hour to file an international flight plan and, once approved, you had one hour to be off the ground. If you didn't make your scheduled time you had to start over. Third, the airplane had to be inspected to make sure it was ready for the Ambassador and the flight. Arriving at the plane some forty five minutes before the Ambassador's scheduled arrival the exterior inspection of the plane was completed. On opening the passenger door we suddenly realized the Ambassador was already in the airplane awaiting our arrival. When our apologies were offered the Ambassador told us to relax and take our time as he had come to the plane early where he had been enjoying the chance to sit and think in peace and quiet. He then told us to depart when we were ready. Such a relaxed, easy going manner was totally unexpected. Later we learned that it was his way. We need more high officials with his ability and demeanor.

BRUCE STOCKDELL

THE HAZARDOUS

MIGS

In their Zone the Soviets had three operational airfields two of which were near the corridors. At times they would bring tactical aircraft or other equipment of interest to these fields. Access on the ground was very difficult so flights were made through the corridor nearest the targeted field and an effort made to get information from the air. To do this a Leica camera with a very long lens would be used. These missions were flown in clothing purchased on the local market, the absolute minimum of identification was carried and a parachute worn in case we had to leave the aircraft or go down with it. The flight was made with the camera laying in the lap, cocked, and ready to snap a picture. Flying these missions were the world's worst navigators as it was almost impossible to stay within the boundaries of the flight corridors??? One such mission was assigned because the Soviets had moved elements of a different jet tactical fighter wing into one of the fields. We needed to know what wing, what model of the aircraft it was and, if possible, what their mission was in Austria. Using our usual good navigation my flight came very close (too close) to the Russian field and several excellent photographs had been taken when a flurry of activity was observed. Obviously they had become suspicious and were scrambling a MIG jet fighter to come up and take a look. What to do? With a top speed of about seventy plus miles per hour and a stalling speed of about forty miles per hour against his stalling speed of about one hundred fifty miles per hour (below which he could not stay in the air) he obviously could not be outrun. The options were to play innocent, holding flight attitude flying straight and level gradually correcting back to the corridor, while waiting to learn his intentions or to bail out of the aircraft. There was no desire to do the latter so checking the camera to make sure it was ready a period of waiting and watching began. As he closed in from the rear, the flaps were dropped suddenly reducing the airspeed to fifty miles per hour as he passed about fifty feet to the right going one hundred miles per hour faster. As soon as he passed the camera was raised and one of the best pictures possible captured the entire aircraft with the Unit designation and even the tail number clearly showing. That picture satisfied all of our immediate requirements except their mission. The MIG pilot, apparently satisfied, went his way and did not come back.

Mission accomplished.

HOT CARGO

Another clandestine use for the small aircraft was to transport cargo that was too sensitive, or "hot", to be taken out of Vienna on the ground through the Soviet Zone. This cargo could be items of equipment or other materiel or (more often) it would be human beings. This cargo would be flown out to the American or British Zone (the French Zone did not border on the Soviet Zone). The human cargo would usually be defectors or deserters from the Eastern Bloc.

When it was human cargo questions were never asked for, if caught, it would be better to be ignorant of the identity or background of the passenger. It did make for unusual and sometimes uncomfortable situations. Flying one hundred or more miles behind the Iron Curtain hauling a wanted individual in a two place unarmed aircraft and we not even know who the passenger was or what they had done. Maybe, so, ignorance is bliss????

One day a call was received relaying instructions to have a plane at a given location at a specified time to pick up a man and heavy emphasis was placed that no questions were to be asked of this passenger. He was to be taken to a specified drop point in the American Zone then the plane was to get out of the area as quickly as possible. When reaching the pickup point it was very clear that the "cargo" was someone special because of the way he was received and handled but no questions were asked and no further information given. He was delivered to the specified point then the plane returned home. Several days later in our American movie theater a newsreel was shown. The feature article was this "hot cargo" being received at the White House by the President of the United States. He was a Minister of the Hungarian Government who had defected to the West. He and his wife (who was an American citizen) had defected across the border between Hungary and Austria. In the process she stepped on a land mine and lost a leg. She was treated for her injuries and later made it to the US. We were not playing games.

A NEAR MISS

Flying small planes in the environment existing in Vienna behind the Iron Curtain was very hazardous in many ways. First, the weather, especially during the winter months, could be severe and present many difficulties. At best there was very little weather information and what was available was not reliable.

Flying aids were essentially nonexistent. For the ten mile wide corridor that was the one hundred mile primary access route between Vienna and the American Zone there was only an old style radio range, which was weak and not very reliable at one end and a weak unreliable radio beacon at the other.

A radio range of that vintage transmitted signals in four quadrants with opposing quadrants (northeast-southwest, for example) sending an N (DAH-DIT) in Morse code and the remaining two quadrants sending an A (DIT-DAH). It also transmitted a periodic identifying signal which allowed the pilot to identify the station. Where the quadrant circles joined, the N and the A overlapped to give a solid tone. The pilot listened in his headphones and flew the plane as necessary to maintain a constant tone. All is well and good if the tones received are loud, clear and reliable. These were not.

A radio beacon, of that vintage, sent out a constant radio signal with a periodic identifier. The aircraft was equipped with a receiver tied to an instrument containing a directional indicator needle which pointed to the station. The pilot flew to keep the needle on zero which represented the station. All went well if the signal was loud, clear and reliable. This one was not.

The aircraft being used had the absolute minimum of the most primitive instruments to aid the pilot when flying in inclement weather. In reality, the instruments were essentially the same as those used by Lindbergh when he flew across the Atlantic Ocean in 1927 with one exception, the radio direction finder for the beacon.

One day the author started for the American Zone. The weather looked good and there had been no reports of bad weather for the route. About half way through the corridor, flying over low lying mountains, clouds were encountered which did not appear to be heavy enough to cause concern. A decision was made to climb above them then drop down on the other side of the mountains and continue on course. However, as the flight proceeded the cloud tops kept rising to above twelve thousand feet, the maximum safe altitude without oxygen (which wasn't aboard), so a decision to reverse course was made but then it was learned that the clouds had closed in behind and the plane was flying blind "in the soup" without the necessary cockpit instruments.

That day the radio aids were not working well so all that was left was dead-reckoning navigation, using the needle, ball, airspeed indicator, compass, rate of climb indicator and altimeter. For straight and level flight, with these, the needle is centered, the ball (which operates like the bubble on a carpenters level) is centered, the altimeter is steady, the rate of climb is on zero, the airspeed Indicator shows the speed you have set with the throttles and the compass indicates the course selected. The flight was in trouble and a decision was made to set a course that would return to the departure point then start a 500 foot per minute controlled rate of descent that would, hopefully, bring the flight out of the bottom of the clouds on a course to the point of origin. All that was needed was to maintain a straight course, keep the wings level, maintain a steady rate of descent, keep the engine from developing carburetor icing (from the moisture in the clouds) and pray. A lot of praying was done.

All seemed to be going very well until it was realized that the needle was to one side, the ball was to the other side and the vertical speed indicator was unwinding much faster than the desired 500 feet per minute. Strange, for it felt that everything was as it should be. The author had logged many hours of instrument flying and had instructed instrument flying. Now that experience and discipline began paying off. It was a classic case of "vertigo", the plane was in a very steep spiral going down faster and faster. Vertigo is a condition in the middle ear which renders your personal balance system completely unreliable. (This is almost assuredly what happened to John F. Kennedy, Jr in July 1999). It is scary. As soon as the situation was recognized corrective action was started, step by step, ignoring personal feelings, to center the needle and the ball and pull out of the dive.

As corrective action was finished the plane broke out of the clouds in a dive over a tree-covered mountain missing the tree tops by about fifty feet. The rest of the trip was uneventful on the return to Vienna at tree top level under the clouds. Until this happened, there had been skepticism about vertigo-no more. We learn the hard way. A very close call indeed-lucky to be alive.

THE SUDDEN STOP

One of the most used "air strips" was about 1400 feet of a city street that was alongside the Danube Canal. The street curved about ten to fifteen degrees and had a forty foot high bridge at each end. The very cold and swift canal ran alongside the outside of the curve and a row of two-story concrete houses were along the inside. The only landing assistance was a windsock on top of a nearby building. It gave the wind direction and the "angle-of-the-dangle" of that sock in the wind also allowed a fair estimation of the strength of the wind. For a safe landing a maximum safe wind speed of thirteen to fifteen mph had been set. A resident, not a flier, was responsible for caring for that sock and having it out when needed.

Starting though the corridor alone with the back seat full of "hot" materiel a problem dictated a turn back. The windsock was checked to determine the direction for landing, and the estimated wind speed was barely within the maximum. Proceeding to land all was going well until the plane was about three feet off the ground and ready to touch down (the pilot has the least control over the airplane at this point) a gust of wind came roaring between two buildings, turned the plane ninety degrees and it flew straight into the side of a large multi-story concrete house. The airplane was demolished but fortunately the pilot was not hurt.

Flying in civilian clothing in the Russian Zone with a load of "hot" cargo was going to attract a lot of unwanted attention. To avoid that the author slipped out of the wreckage, hurried between the houses and made his way around approximately two blocks of houses then joined the back of the crowd that had gathered around the wrecked plane. Everybody was asking, "Where's the pilot?" Tass, the Russian News Agency was photographing everything. The author, passed as one of crowd, and no unwanted identification was made.

A later investigation revealed that the windsock had been changed that day, from a summer unit to a much heavier winter unit, which would change the readings from fifteen to closer to twenty five mph-much too high for landing. No one had advised the pilots. Had this information been availablethere would have been no attempt to land.

The investigation also showed that a landing gear hydraulic strut had snapped off and flipped up through the side of the plane directly beneath the pilots knees as the legs were bent in the normal position while flying this airplane. It could not have missed by more than one inch. If it had hit a leg would have been lost. Again, someone was looking over his shoulder.

After any aircraft accident, including one like this, the pilot must pass a flight check before he can resume normal flight operations. As the check was successfully completed the check pilot said he had heard much about that strip in flight circles and asked if it would be possible to see the scene of the accident. After flying through the corridor to the location he was shown how the sock was checked the plane flown perpendicular across the center of the strip then a right or left turn executed to land in the direction the sock had indicated. That day was very choppy day and the plane was bouncing around like a rubber ball. After flying a normal approach pattern, dropping down over the bridge, flying the length of the curved strip then up over the bridge at the other end the check pilot was asked if he wanted to go around and come in for a landing. With his voice a little shaky he said "No sir, if you have been landing on that strip you have passed the test".

END OF CHECK RIDE

The pilot must review the accident report and sign it. Then it goes through the chain of command to be to be reviewed and endorsed. Much later the final report with endorsements was reviewed and it was learned that one Commander had endorsed it as follows: "I am personally familiar with that landing location, having flown off of it myself, and it is my considered opinion that the Air Force Distinguished Flying Cross (DFC) should be awarded for each take off and each landing made there" then signed it.

I never did get my DFC???

THE WORST DAY

It was a Saturday. There were three officers assigned to that technical intelligence office at that time each specializing in a different technical area. The office was manned on a standard five day work week with one officer pulling a Saturday morning standby shift to handle any unexpected business. This was the authors day for duty and the other two decided to make a corridor photo run.

At ten in the morning a call was received advising that our plane had crashed and there were apparently no survivors. As the only Air Force Officer left in town all necessary actions fell on the author. These included confirming there had been an accident; confirming it was our plane; identifying the bodies; notifying the wives and/or families; and finally retrieving the camera equipment and the enclosed film. Next was locating and notifying the appropriate people and offices up the chain of command. Later this included arranging for the shipping of the bodies and closing out the two families and getting them and their possessions on the way back to the USA.

The film contained good, high quality pictures which were very valuable to us but oh what a high price we paid. The Air Force lost two superior young officers who were outstanding intelligent collection specialists.

One of the young officers was a hero from World War II. He had flown a badly shot up Boeing B-17 Flying Fortress bomber from over the continent to England by himself when all others on board were incapacitated. For this he was awarded the Distinguished Service Cross, our nations' second highest combat award at that time. Two very close friends were lost.

This work and the risks taken received very little attention and not much support but it was very valuable, and very important to our country. One of the reasons these stories are being written is so that these men and that work will not be forgotten.

This book is dedicated to the memory of two very good friends who gave their lives that we might continue to enjoy our freedom. I salute you Fred and Mal.

REST IN PEACE

THE HUKS

Several years after the preceding accounts, during the Viet Nam War, the author was Director of Operations for the 13th Air Force DCS/Intelligence with Headquarters at Clark AFB in the Philippines with responsibilities covering about one-fourth of the globe including almost all of Southeast Asia, except Viet Nam. That was the responsibility of 7th Air Force.

Inside the Philippines were several areas of unrest including the Moslems areas of Mindanao and the insurgent Communists known as the HUKS in Luzon. In addition to other responsibilities we had to watch the two groups and keep our government informed.

One day while in the air a radio transmission was received asking that we divert to a specified location to pick up a passenger. The manner and tone of the transmission indicated that we were not to ask questions. Proceeding immediately to the location the plane was met by a heavily armed combat team in jungle battle dress escorting a badly wounded man, bandaged from the waist up. The engines were not shut down as they loaded him aboard and one of the escorts came on board to advise where to take the man and his escort. The passengers were to be dropped then the plane proceed on its way with the crew forgetting what we had seen. This was done.

Returning to the office and relating to our bosses what had happened raised much interest in the staff. The next day it was learned there had been a strong clash between the HUKS and one of the counterinsurgency groups. One of the leaders had been badly wounded. He was the passenger.

There have been many unusual and very interesting flying experiences including this one.

TEST FLIGHT

While on the command Staff in the Philippines, there were several additional duties. Among those was flying the Commanding General on local trips in very nice C-47 aircraft assigned to him. Besides his plane there were several other C-47's located at the Base. Since the author had more flying time in the C-47 than any other assigned pilot he was also designated an instructor pilot and the maintenance test pilot for that type aircraft. After any major maintenance is performed, especially that which might affect performance or safety, the plane must be test flown and approved by a qualified test pilot. During these flights the repair must be tried to the maximum to see if it will perform properly or break. Those flights are not for passengers.

One day he was called to test the General's plane. If it passed, he was to take it on a mission to pick up passengers and return to Clark Air Force Base. Normally, at the end of a test flight the plane lands and the test pilot fills out and signs the test report which releases the plane back to line duty. On this particular occasion the left propeller had been replaced after complaints from an earlier mission. It seemed routine and, since the test crew was also to be the mission crew, it was elected to call in to report successful completion of the test then proceed on the mission intending to sign the papers on return thereby saving more than an hour. This plan was approved.

When the crew chief learned of this he showed obvious concern. Long experience teaches that you always pay attention to and listen to your crew chief and this one was especially respected as numerous flights had been made with him. When asked what was bothering him he said he didn't think the propeller was the cause of the problem reported and asked that we not to go on the mission without first landing, following the flight test, so he could make a visual inspection of the plane before we proceeded. This airplane was his 'baby" and he practically lived with it so his request was heeded and we reverted to the original plan to land after the test.

A normal, routine flight test was completed including putting maximum possible stress on the newly installed propeller. Everything performed exactly as desired so the test was terminated, the control tower notified and the plane headed for the airfield. Normally the engine controls would have been placed in cruise position but being close to the downwind entry point of the traffic pattern they were left in the climb position which gave maximum power. Later we were glad they were. All was going well as the two pilots and the crew chief, who sat on a three legged stool between the pilot and the copilot, discussed what might have caused the problem reported on the earlier flight.

The plane was barely established on the downwind leg when there was a loud "BANG" and the airplane began to shake so violently that it was difficult to keep in the air. The impact was so strong it threw the crew chief off his stool onto the floor in the doorway between the flight deck and the passenger compartment. A look out the left window revealed the source of the trouble. The newly installed propeller on the number one engine was hanging down at an angle, still under power and trying to pull the plane down on the left side. "Feather #1". With this order the copilot is supposed to immediately start a set emergency procedure to feather the propeller and shut down the engine thereby reducing the aerodynamic drag on that side which would help keep us in the air. At this point the copilot froze and, with the crew chief flat on his back in the aisle, there was no help. We were in serious trouble.

Over the pilot's head, in this plane, are two red buttons positioned close to each other, one for feathering each engine but other actions are supposed to precede hitting those buttons, for reasons of safety. But, if we were to survive, there was no choice. Slapping the left red button, bypassing the safety measures, caused the left propeller to begin to feather immediately, and the engine to shut down and vibrations to reduce to the point that the airplane could be brought back under control. By this time the crew chief had managed to get back to his position and he finished the shutdown procedure for the left side of the airplane.

Finally the copilot recovered enough to call in an emergency which was acknowledged by the control tower. Continuing on the downwind leg to the only runway (11,500 ft long) we were advised from the tower that we must extend our downwind leg as we would be #3 on the final approach to landing. Not believing what was heard because a declared emergency automatically moves a plane to #1 for landing the author took the microphone and reminded the tower operator of that fact. His reply, "Please sir, you will be the third emergency on final approach. The first is a C-141 transpacific flight inbound from the United States with an engine fire and I've directed him to land long then roll off the runway at the end; the second, a KC-135 flying tanker has lost his hydraulic system and will land short then turn off the runway at the middle (when that type plane loses their hydraulic-system the refueling boom acts like "the tail wagging the dog" making directional control of the aircraft very difficult so their need could not be disputed) and you as the slowest will be third and I ask you to land short then roll off as soon as you can. A good plan.

There were five turnoffs/taxi ways for the two mile long runway and each was crowded with emergency equipment. We hoped they would not have to use the equipment but it was comforting to see it there. All three aircraft landed without problems and everyone concerned breathed a big sigh of relief.

On climbing out of our plane we saw a very unusual sight, the left propeller pointed down instead of straight forward. We later learned we had a broken crankshaft in the engine and it had separated. This was the only incident of it's kind that this pilot had heard of in all his years of flying.

The first action on entering the flight operations building was to recommend that the copilot be brought before a Flight Evaluation Board to determine his fitness to remain on flying status. You cannot have a pilot who freezes during an emergency. Suppose he had been the Flight Commander?????

THE ZAMBOANGA STORY

Serving as Chief of Intelligence Operations for the US 13th Air Force in the Philippines included serving on the US country intelligence team. The team was composed of representatives of every US intelligence entity present in the country including the Army Navy, Air Force, CIA, FBI and others. The job was to keep informed on all activities in and around the area to help keep our country out of trouble. This also included being the Air Force member of the joint US-Philippine Intelligence Council. This Council consisted of the US Army, Navy and Air Force and the Philippine Army, Navy, Air Force and Constabulary. The Council met on a quarterly basis to keep the two countries informed and coordinated in areas of mutual concern.

Hosting the Council meetings was rotated among the member organizations. When the Philippines hosted the meetings they were held in different locations around their country to better acquaint US representatives with their country as well as local problems and capabilities. One such meeting was scheduled for Zamboanga the biggest city on the large island of Mindanao which is also the southernmost of the seventeen hundred islands which make up the Philippines. (Zamboanga is just north of the equator and is one place where black pearls are found. It was also the subject of one of Bob Hope and Bing Crosby's movies called "On the Road to Zamboanga"). When you stand on the shore in Zamboanga you can look across a narrow strait and see the island of Borneo, which is a part of Indonesia. This location is six hundred miles south of Manila and even further from our home location. Transportation was scarce. To solve that problem our leaders offered the use of one of our planes, a C-47 to transport the group to Zamboanga. Since every seat was needed to carry the entire group including the author he was assigned to pilot the plane, participate in the meetings, then fly the group home.

Taking off, fully loaded, from the Manila International Airport the thought hit, on board this aircraft were all of the top intelligence people of the US Army, Navy and Air Force and the Philippine Army, Navy, Air Force and Constabulary. If anything happened the entire intelligence capability of both governments in the Philippines would be wiped out! What a responsibility.

Luzon, our home island, was the northernmost and Zamboanga is on Mindanao, the southernmost, of the islands of the Philippines. The six hundred mile flight from Manila to Zamboanga is very beautiful with seventeen hundred islands scattered over a deep blue sea. At that time modern airway control facilities extended only part of that distance. That meant the last two hundred miles had to be flown by the old "seat of the pants" flight rules. Complicating the matter was a low level cloud cover over the last portion of our trip and our destination.

The only aid available was a simple radio beacon on the Zamboanga Airport but its location behind a nearby mountain made it unusable for any distance more than fifty miles out from the airport. To meet these conditions we homed in on and flew over the beacon above the clouds at 10,000 ft, past the peninsula to an open area of the sea where we found a break in the clouds then steeply spiraled down to come out below the clouds at five hundred feet above the water. From that position we could see our destination and flew directly to the airport where we landed to complete our mission. We would not normally do that with passengers aboard but our only alternative was to return the group to Manila. After the flight the senior US officer, a Navigator, commented he hadn't been in a maneuver like that since World War II. When advised of our choices he understood and agreed.

As the plane taxied in to the terminal it appeared they had spruced up the facility in anticipation of our arrival. We were guided in to a newly paved parking area. While looking out the Pilot's window the crew saw the left wheel start to sink into the new pavement. To prevent getting stuck power was immediately applied to move the plane back to some of the older more solid pavement. On leaving the plane we could see two rather deep dimples in the pavement of their new parking ramp. The new pavement seemed to have been laid on an inadequate base. We almost became stuck in Zamboanga.

During the flight a message had been received advising that the plane was needed for another mission, changing our flight orders to dropping the passengers and immediately returning the plane to Clark AFB. Another flight was laid on to pick them up one week later. The author also flew that trip but missed the meeting.

After a very good lunch provided by our host the plane departed. It turned out that the return was good for a news report came out that US and Philippino officers had assembled in Zamboanga to plan military hostilities in the area. Presence of the US plane was given as evidence. Of course there was no truth to the story but getting the airplane out of the area of Zamboanga helped diffuse that rumor.

IMMINENT HOSTILITIES??????

Several months after the preceding story, there was another incident involving Zamboanga and Mindanao.

Through unsubstantiated sources word was received that a force was massing on Borneo for an invasion of the Philippines through Mindanao. Not believing this report because of our knowledge of the area and its' people and because we also knew that neither side had the means (financial or materiel) nor the will to mount and maintain such an attack. But we were obligated to check out the report and forward the information to our Headquarters.

F-4 reconnaissance planes flying in the area were tasked to fly over the subject area and report their findings. As expected they reported no such activity and no massing of forces of any kind.

Since it was mandatory that we report any "Imminent Hostilities" and such reports are automatically classified Top Secret a report was prepared and forwarded through appropriate channels. The report gave all facts, along with the conclusion that it was baseless-no reason for concern. The DCS/Intelligence was away so his deputy signed it and the Commander (a three star) signed off so it could be forwarded.

Several nights later, late in the evening, an urgent call was received about an incoming message concerning "Imminent Hostilities". The author replied that he was familiar with the subject as he had written the report in question. This and the fact that all file copies of such documents were under control of the Operations Division led to orders to get to Headquarters ASAP. En route, after picking up the Sergeant (NCOIC), we walked into a hornet's nest.

At the highest levels someone apparently did not read the whole document and panicked at the "Imminent Hostilities" label. Four Star Generals were called out in the middle of the night and tempers were boiling. No one at that level apparently knew anything about the subject. The Colonel (Ass't DCS/I) who had signed the document and the Commander (3 stars) who had approved it were both away and neither the Deputy Commander (two stars) nor the DCS/I had knowledge of the documents.

Retrieving our file copy of the report it was given to the DCS/I, along with supporting facts, to take it to the Deputy Commander's office. This was done and the DCS/I returned in a few minutes saying that there was apparently no problem and that we should secure the documents and go home which we did.

As he walked in the front door the telephone was ringing. Another four star General had been called out and we were to get back to the office ASAP. This time the four star on the telephone had chewed out the two star and he was boiling.

Both he and the DCS/I had very short fuses and they clashed. The DCS/I was fired on the spot, transferred back to the USA and immediately retired.

Subsequent reviews proved that everything had been done exactly according to the book but someone "topside" had hit the panic button. The end result was that a good friend and a good boss were lost while the Air Force prematurely lost a good senior Intelligence officer.

IT WAS ALL A BIG BROU-HA-HA ABOUT NOTHING.

ELECTRONICS

ELECTRONICS

One of the most critical technological fields in modern times is electronics. Electronic applications are in every phase of our society and in every piece of equipment we use. This includes defense uses in aircraft, computers, communications, test equipment and almost everything else.

Each type of equipment sought required different sources and different collection approaches. Sometimes it required quite diverse operations so we tried to use every channel and resource. Some of these are presented in the following pages.

FIELD COMMUNICATIONS

Quite often those in the field had opportunities to collect equipment for other agencies or organizations which could significantly enhance the overall capability of the United States. This always gave great pleasure, partly because of the friendly competition between agencies, but, more important, because it made us feel more valuable to our country.

One classic example, from the height of the Cold War days, was a request from a top communications security Agency for help in gaining access to Warsaw Pact field communications by acquiring some of their equipment. At that time all Warsaw Pact powers, including the Soviet Union used East German equipment which was equal to, or better than, any other communications equipment anywhere in the world. We were eager to get our hands on some of this equipment for a number of reasons, but this particular requirement pushed it far up the priority scale.

Our best people were chosen to go after this equipment. Before long before we had our first complete set consisting of two end terminals plus the antennae and all peripheral gear which was immediately put into use. The equipment operated on the line-of-sight principal, so all the field operators had to do was find an operating location that placed them on line with the targeted communications terminal (s). Since our equipment was not only compatible, but identical in every respect to the targeted Warsaw Pact equipment, their privileged communications could be read without difficulty. Eventually we acquired numerous complete sets allowing all Warsaw Pact communications to be covered while at the same time permitting exhaustive analysis of the excellent equipment.

Triumphs like this, out of necessity, were known to very few people but brought great satisfaction to field collectors. We often hear about the failures of our intelligence people but we rarely hear about the resounding successes such as this. Now we can tell about some of the successes achieved, many times at great risk to those involved. We owe a great debt of gratitude to these people who worked so effectively behind the scenes.

ANALOG COMPUTER

Early in the program, before the days of modern computers, we learned of an Analog computer developed in Czechoslovakia. It was reported as "representing the State of the Art" at that time in Czechoslovakia so immediate action was initiated to acquire one.

On its arrival we learned that the equipment was quite primitive by our standards but it did work. No usable technical data was gained from this "State of the Art" acquisition.

What was learned was that the machine was constructed primarily of parts "Made in the USA". Almost all of the gauges and other control parts were on the US embargo list and should not have been available to them!!!

Obviously their Foreign Materiel Acquisition Program was working well.

MINIATURE TUBES

When the "commercial" business first started we would often ask for specifications and other data on equipment of interest. Some times this data alone would satisfy our needs so we didn't" place orders for the equipment as would normally be expected from a profit making business. This was done so many times that our bosses were finally told we had to place some substantial orders to establish ourselves as serious business people.

At that time the Soviet miniature electronic tubes appeared to be of good quality but we had never taken a close look at them. Since we had no technical need for the items it was decided to order 1000 each of ten different Soviet miniature tubes, for a total of ten thousand tubes just to meet our operational need for a large commercial order. It was placed and immediately filled.

When the ten thousand tubes arrived they were sent to the Air Force laboratories for exhaustive study, tests and analysis. The results surprised all concerned and it made this acquisition very valuable. The tubes were found to have four unique features which were judged superior to competitive US tubes. One was their way of cutting the mica wafers inside the tubes. Another was their method of closing the glass envelope after evacuation of the air. Their method turned out to be easier, better and cheaper than ours. As soon as we learned the ways to make a "better mouse trap" the information was forwarded to US manufacturers as quickly as possible without revealing where the information came from nor how we got it. The laboratories issued bulletins to the manufacturers giving "new methods developed in the labs which might help them make better products".

All US manufacturers immediately incorporated the four improvements into their production tubes. Ironically, prices quoted by all manufacturers to the Air Force for it's next acquisition of tubes were increased to reflect the cost of the improvements incorporated. The Air Force had to pay twice for these "improvements" but we did get a better product to use.

Such is life!!!!!!

TRANSISTORS

While monitoring technological developments we learned of a new Soviet method of making transistors for use in computers and other high technology electronic equipment. We immediately activated our collection mechanism and some of the transistors were acquired for analysis.

Early the next week a visit was made to the office of the President of the Federal Systems Division of one of our leading Computer manufacturers to ask if they would be interested in analyzing the items under contract and, if so would he submit an analysis plan and cost proposal. He said they would and asked if he could have the items to study and help them prepare a better plan. The proposal was promised on the following Monday.

On Monday his office advised that he had to leave town and would be back on Wednesday. A visit was made on Wednesday to his office expecting to receive their proposal. Instead he returned the transistors along with a finished analytical report giving the characteristics, performance, and the unique manufacturing techniques used in making the items. When asked how he had accomplished that so quickly, he replied that on receiving the transistors he immediately assembled, at a remote laboratory, a team of top experts from throughout their company to study the items. They had worked twenty four hours a day analyzing and testing the transistors then writing the report. He handed me a copy of the report which had been finished late the night before.

When asked what the cost would be he answered there was no cost for this information was also valuable to them. When reminded that they had no proprietary rights to the information since we had provided it to them he said he knew that but this action gave his company a six months lead on his competitors. That is the way big business operates.

This was unorthodox but it gave us the information much faster and was immediately made available to all. Altogether it was a very good operation.

CATHODE RAY TUBE

One stage of the undercover career involved running a civilian export-import company in a major city in this country. The firm conducted regular import/export business in eighty-eight countries and we managed to bury our special interests in their volume of business to avoid unwanted attention. Every effort was made to keep our special shipments away from the office by routing them through the company's freight forwarding firm.

On one occasion this didn't work. A box covered with Cyrillic lettering was delivered to the office. It caused a lot of discussion among the staff members. One elderly gentleman, a valued employee, was especially concerned. He asked to speak privately to express his deep concern as a good, strong, concerned American citizen about trading with the Russians. It was explained to him that, even though our country was farther advanced in most technological fields we did not have a monopoly on brains or good ideas. Other countries could, and did, come up with ideas that we hadn't yet thought of and developed manufacturing methods that are superior to ours. Then I pointed out that American companies needed to know these things so they could meet or counteract the foreign technology to better compete in the world market place. He was told that one of the easiest ways of gaining the needed information was to buy their products for study and analysis and this was what we were doing.

The gentleman considered this, then said he had not thought of it that way and that it made sense to him. Everything told him was true. The only thing he couldn't be told was that the customer was the United States Government. Regrettably he could not be told the full story. This old gentlemen and his concerns were very highly respected.

RUBY LASER ROD

In the early days of laser technology there was great interest in laser developments throughout the world. During this period we learned of the development in a friendly foreign country of a new synthetic ruby laser rod of extraordinary purity and clarity. Our laboratories were very interested in getting their hands on one of the rods for study and testing but for some reason did not want to use normal procurement channels. Instead a decision was made to use our system. It proved to be one of our easier assignments for, in a very short time the rod was in hand.

The rod was about twelve inches long and 3/4 inches in diameter, deep red in color, had no visible flaws and seemed to be of jewelers quality. It was a thing of beauty and getting to see it was a welcome fringe benefit.

BRUCE STOCKDELL

NUCLEAR

NUCLEAR

Another technical field demanding special attention was the nuclear realm. Anything "nuclear" drew immediate response whether it be weapons, research, power generation or other things such as materials, written articles, etc: In the early years, the word "nuclear" was magic and received immediate and undivided attention.

Anything in this field was also very difficult to obtain. However, we did have some successes as reported in the following.

DUOPLASMATRON

In the early days of "Star wars" thinking, there was intense competition between two exotic weapons concepts. One group was promoting Laser weapons and another Charged Particle Beam weapons systems.

While walking down the hall one day in the building that housed both weapons groups, an approach was made by a young Lieutenant with the reputation of being a real "Brain". and one of the chief proponents of the Charged Particle Beam system. He said he understood that unusual equipment could occasionally be obtained from "special" sources. When that was confirmed he was asked what he needed. He asked if either a Unoplasmatron or a Duoplasmatron could be obtained. In reply he was told that if he would tell us what they were and what they were used for we might be able to give him an answer. None of the collection group had ever heard of either.

He replied that our charged particle beam research required a strong, steady flow of negatively charged ions and we didn't have equipment in this country that could meet this need. Also other intelligence information indicated that very special equipment for this purpose had apparently been developed by interned German scientists working in a Soviet laboratory. He was able to provide the names of the German internees along with some very helpful technical data. The first unit they had developed was called the "Unoplasmatron" and the more advanced one the "Duoplasmatron". Either one would satisfy this need but getting either presented problems for collection people.

After further discussions our investigation began and our collection plans were laid. We learned the leading German scientist on this project had just been released from internment and returned to his home in Eastern Europe. Further checking revealed he had opened his own laboratory there and, in all probability, was continuing his highly specialized work. Proceeding upon this assumption a high priority collection effort was mounted which succeeded far beyond our wildest expectations. The scientist was badly in need of money so he was very cooperative and open to an opportunity to make money. In short order there were two of the very strange looking Duoplasmatrons in our possession. Without delay both were sent to the National Laboratory which had the greatest need. There the equipment was immediately put on line and evaluated as it was producing. In a very short time thereafter, the Laboratory Director informed that installation of the two items set their research program ahead two years.

Altogether twenty of these units were obtained for our research institutions. A tremendous success story.

THE RESEARCH LABORATORY

In earlier days anything with the word nuclear attached immediately received attention and a very high priority in the collection arena. Information in this area was also very difficult to obtain.

At one time we knew the Soviet Union had three Nuclear research laboratories established and operating in the Eastern Bloc of nations as we had good intelligence on these facilities, their personnel and the work they were doing. However, there were some missing links. Assessment of their operations led us to believe there must be a fourth laboratory somewhere which would provide the missing pieces. We didn't know where it was, what they were working on or who was running it. Getting this intelligence became one of our highest priority collection requirements.

About this time we also learned the Soviets were making a strong effort to obtain a sizable quantity of optically pure borax. Borax of the quality sought was, at that time, only used for two purposes. The first was to produce very high quality optical lens and other optical equipment which had to meet very rigid performance standards. The second was as a catalyst in some types of nuclear research. Either usage was of great interest.

The quality of borax needed was only available in one place in the world, the mines located in Boron, California. It was considered a very valuable war commodity and was under very tight control.

We identified their procurement mechanism and learned the quantity sought was a minimum of one hundred tons. After careful study, it was determined that the amount sought would not substantially advance their capability in either field, if it got through. But, if we allowed them to procure the borax, then we could track it to its destination which might provide us with very valuable intelligence about their nuclear research capabilities. After deciding it was worth the cost we recommended and received approval to mount an operation that would facilitate the Soviet procurement of the desired one hundred tons of borax. The plan was that we would monitor the procurement then follow the shipment to its' destination. This way we would know exactly where it was going and how it was to be used.

To help us at this time we had a remarkable group of individuals in the Eastern Bloc of nations. Certain local citizens had the ability to remember long, complex numbers and their eyesight was so sharp that they could stand by the railroad tracks as a freight train rushed by and tell what freight cars, by number, made up that train. This capability was very valuable.

In subtle ways we made sure the Soviet procurement efforts were successful and that the shipment went from west to east through the Iron Curtain without excessive difficulty. The numbers of the box cars containing the borax were then passed to this network of trackers in the East. They tracked the cars to their destination, reporting back as they passed through each terminal, check point or crossing. As hoped, the destination was the long sought after fourth nuclear laboratory. We were then able to successfully direct collection efforts against the lucrative target. This simple operation provided great gain for the United States at minimal cost.

Two or three years later a senior intelligence officer was transferred into my work area. As we became acquainted, and exchanged war stories, he told that he had worked in the office that was responsibile for preventing contraband shipments going from west to east through the Iron Curtain. When asked how successful they were he replied that they were very successful except for one shipment which defied all their efforts. When asked if he knew what the shipment contained he replied it was one hundred tons of optically pure borax. When the author started to laugh he wanted to know what was so funny and wasn't amused when told we had helped make that shipment. He calmed down when told why, and what had been accomplished. However, he thought they should have been informed. Although we had nothing to do with that decision, it was probably a wise one, otherwise they might not have made their normal efforts to prevent it, thereby alerting Soviet intelligence that "not everything was as it appeared." It proved to be a very successful operation, and important information was gained for our country.

TRANSLATION POOL

Throughout the world and, especially in the USA, a lot of valuable scientific, technical and developmental information is published in unclassified journals, periodicals and reference volumes readily available on the open market.

There are two trains of thought in the US on this practice. One is; we must publish information on our technological advances for the information and education of our own engineers and scientists so that we don't waste our brainpower and energies "reinventing the wheel". Of course there is some valid basis for this position since we do need to keep our own people current on the state of the art. To a large extent this was proven with the publication of a seven volume encyclopedia on electronics, published by one of our major publishing companies during the early years of development of modern electronics. Publishing was justified by claiming that our own specialists needed this information readily available to them. This set is credited to a large extent with accelerating the very rapid development of our national electronic mastery.

The other position is that just by digesting the contents of this set, any country was able to make a quantum ten year jump in their capability, making our latest developments and knowledge available in such readily accessible easily understood form to the general public. For the cost of a subscription was, in essence, giving away our top secrets. All a foreign power had to do was buy a subscription or the set of encyclopedia. Many believe we should do all we can to make it more difficult for them. Both positions have merit.

While we publish far more information we are not the only country to publish our secrets in this way. We decided to try to capitalize on this fact by subscribing, through diverse channels, to a total of 150 different Trade, Scientific & Technical Journals and Engineering periodicals, magazines, and other publications from targeted countries.

Our location, at the time, was ideal for this activity since most of the people living there are multi-lingual and many have technical backgrounds. Many were also newly released prisoners of war and we could be employed at one tenth the cost of such services in the USA. A translation pool averaging twelve to fifteen qualified translators was set up to translate the tables of contents of the publications. These would be reviewed and selected articles which seemed to be of special interest. Condensed translations of these articles would then be made and forwarded with the publication, through channels, for further consideration.

The man in charge was a very interesting study in humanity. When World War II started he was drafted from Engineering school into the German Army. As the war neared an end he was stationed in a Russian controlled area of Eastern Czechoslovakia. The Germans were obviously losing and he, not wanting to surrender to the Russians, deserted and made his way across Czechoslovakia to an American controlled area where he surrendered to US Forces, thinking he would fare much better.

Then the rude awakening! The unit he had walked away from was in Russian controlled territory. In accordance with agreements made between the USA and the USSR, he had to be turned over to the Russian Forces. Then the Russians proceeded to send him to a prison camp for the next five and one half years!!!

He was able to adjust himself to his plight and to long term imprisonment quickly learning the Russian language then becoming the camp "Stakhanovite" (or camp leader) conducting all negotiations with prison officials in behalf of the prisoners. This, in turn, made his life a little easier and gave some extra privileges. For example, in later years, he would occasionally receive passes to go off the prison grounds and into the local town for an evening or weekend. Because of the location of the camp escape was out of the question but, as he said, "the passes did make imprisonment a bit more tolerable." It also awakened in him an interest in the study of languages. Upon release, he changed his University major from engineering to languages.

When this man came to work he was a bundle of nerves due to his imprisonment, the long interruption in his life and his intense desire to prove himself. He would not even lay down his work to eat lunch. Because he was such a valuable employee, for his own good and concern for his welfare he was directed to leave his work, take his lunch break outside under a tree and relax as he ate. It worked.

This man was fluent in six languages and, in translation work, he could handle nine more for a total of fifteen languages. This was tested one time when he handed in a listing of an article about "Core" physics. Questioned because we thought a mistake had been made, since we did not recognize "Core" physics, he took us through six languages to show where the root word originated to prove his translation was correct. Then it hit: "Core" was "Nuclear". It was very early in the nuclear period and this was one of the first articles that we had seen about Nuclear Research in Eastern Bloc countries. The entire article was translated and turned out to be most valuable.

Learning from this experience we were very careful about challenging his translations thereafter. On our departure from the area he was recommended to the US Embassy as a staff translator. He was employed and, several years later, was still there.

After the way this man had been treated when he surrendered to the US Forces we marveled that he bore no apparent grudges and that he would agree to work for the US Government. He was a good man. Much could be learned from this example.

FORMER PRISONER OF WAR

For several years after World War II Germans and Austrians, held by the Soviet Union as prisoners of war, were periodically released and allowed to return home: if they could find their home or family. Having access to those lists we often had information pointing to where they had been and what they had been doing while imprisoned. Since some were potentially valuable intelligence sources, the lists were screened very carefully.

One prospect was reported to have worked in the Nuclear Research Laboratory which was the subject of an earlier chapter. He was of great interest to us so immediate contact was attempted. The man was very reluctant to talk after his many years in the Soviet prison camp but finally did agree to an initial contact at his family's home in the evening. Two of us went to his home. On arrival we could readily understand his fear. The house sat on a cul-de-sac less than one hundred yards from the Soviet Zone of Austria. The street going into the cul-de-sac narrowed to a single lane, providing a perfect set-up for an ambush. Not liking the setting we proceeded anyway. To reduce the pressure on the subject it was decided best if the other man, an Austrian, made the first contact alone. This would make it Austrian-to-Austrian with no foreigner openly involved.

Staying in the car, as a back up, the author was ready to do whatever the situation might demand. The setting was very dangerous and we felt very uneasy. It did not help when, some twenty minutes after the other man went in, the front door of the house burst open and a man came running out, disappearing into the darkness. A car was never started faster. It started, was put into gear, and the passenger side door unlocked (for the other man) in record time as a fast getaway was prepared. Then—nothing happened! Waiting, anxiously, for what seemed an interminable amount of time—still nothing. Finally, deciding there would be no other activity, the engine was shut down and the author tried to relax.

It was later learned that the "man" rushing out of the house was a grown-up teenager going to join his friends. Of such things heart attacks are made.

We were successful. The targeted man agreed to come into the office two days later for debriefing. All seemed well and we were congratulating ourselves when another problem arose. At the time when the potential source list was received, each entry was usually assigned to a particular organization. This man had apparently not been assigned so we had gone after him as a "Free" target of opportunity. Then we learned that he had been assigned to the US Army. The Army was contacted and told what had been done, then we requested that the Source be transferred to us. The Army refused. We advised them of the special circumstances surrounding this Source and suggested they pick him up when he came in, since he had already committed himself to come to us. They refused, insisting that we back out, and promised they would handle in it their own way. There was no appeal permitted so we had to stand down.

The next day the Army sent two officers in uniform and riding in a Jeep, to the prospect's house in mid-day. The prospect was terrified and literally "ran for his life". He disappeared and his whereabouts were unknown for about six months. Finally he resurfaced in Germany but we never had a chance to talk to him. Not all actions run smoothly.

MISCELLANEOUS

OTHER MUNITIONS ACQUISITIONS

There were many other very valuable acquisitions which do not fit into the categories we have already covered. These items were acquired in many ways and each carries its own fascinating story. We will now tell you about them.

SURFACE TO AIR MISSILE FUSE

During the Viet Nam war, USAF and US Navy aircraft operating over North Viet Nam encountered an enemy anti-aircraft environment far more intense than anything ever before seen in the history of aerial warfare. This resulted in aircraft loss, and damage, and crew loss at rates much higher than anticipated or considered acceptable.

The main bombing targets in North Viet Nam were Hanoi (their capitol city) and Haiphong (their main seaport). The primary weapon used against our planes was the SA-2 (surface to Air) missile which was of a fairly old Soviet design. The missiles were massed around the two targets in concentrations never before seen around any target in any war.

The SA-2 is a long missile launched under radar control. The search radar finds and locks onto the target (a plane) then, at the right time, it automatically fires the missile which remains under the radar's control for the early part of its' flight. When the missile gets close enough to the target for its on board fuse to lock on the missile becomes self-guided as it homes in for the kill.

In order for the missile to operate properly, the radar must be on before, during and for some time after the launch to successfully operate. The missile lifts off its' launcher very slowly then quickly accelerates to super sonic speeds. Because of its length and its slow initial speed the missile cannot maneuver fast enough in the early stage of its flight to keep up with the twists and turns of fighter aircraft on a strike mission. For those two reasons the system is most vulnerable during and immediately after launch.

There was intense activity on both sides, each trying to gain advantage. From the US side two ways were tried hoping to defeat the missile system. One was trying to force the radar to shut down before its normal cycle was complete thereby prematurely cutting off control signals to the missile.

For this purpose the USAF trained a highly specialized group called the "Wild Weasels". Their mission was to fly over enemy territory daring the North Vietnamese to turn on their radars and try to shoot them down with their SA-2's. If the radars turned on the "Wild Weasels" would attack, firing special missiles designed to home in on the radar and destroy it thus shutting down the system. Hopefully this would happen early in the launch cycle. If it didn't the missile would be heading at them.

Second, was having the Weasels and fighter-bomber pilots use violent flight maneuvers to evade the missile as it came toward them. This was only possible if our pilots saw the missile early enough to avoid the warhead from locking on by "Jinking" (rapidly and repeatedly changing the direction and altitude) the plane. Of course this also meant the pilots had to forget about hitting their target.

This whole operation somewhat resembled the Japanese "Kamikaze's" of World War II. If they were a bit late or if they missed the target and the radar didn't shut down they immediately became the target. The pilots flying both methods were considered to be daredevils and heroes.

A means of electronically confusing the missiles' guidance system was badly needed but nothing tried seemed to work. There was a desperate need to get our hands on their equipment to learn why our countermeasures did not work. Finally, we managed to get an SA-2 fuse, still beautifully packed in its' fitted wooden shipping case. It was immediately forwarded to the US Army fuse Laboratory for analysis carrying the highest priority that could be assigned. As soon as the preliminary analysis was completed and a fuse wiring diagram drawn a meeting of top experts was convened. Thirty five of the top US experts, including the project managers of the three countermeasure systems then under development in the USA assembled to learn more about this fuse. The briefer did a masterful job presenting the capabilities and limitations of the beautifully designed fuse. As the briefing proceeded each of the project managers, one by one, informed us that their program had been shot down as the fuse had the capability of defeating every countermeasure system we had, either on the line or under development.

It was a very sobering experience and everyone present left with the knowledge that we had to go back and start over. It was a truly serious matter but it was also highly profitable. It opened our eyes and kept us from wasting our R&D resources by pursuing the wrong goals. A valuable acquisition indeed..

BRUCE STOCKDELL

ZU-23 ANTI AIRCRAFT GUN

Throughout the Viet Nam war, flying low level fighter aircraft missions over enemy territory was very hazardous. The aircraft loss and damage rate due to anti-aircraft gunfire was extremely high. The intense concentration of their guns and their apparent accuracy of fire seemed unbelievable. We couldn't believe they did indeed have the number of guns needed to inflict that much damage. Whether they had that many or not we needed to know how their guns had such accuracy of fire against low flying jet fighter aircraft at near sonic speeds.

World wide collection efforts were laid on to try to obtain one or more of their guns for study. During our search we learned that relations had soured between the Soviet Union and one of its top client states in Africa. Over the years this state had been lavishly supplied with money as well as some of the latest Soviet military equipment. With the break up the flow of both money and equipment dried up. When approached the rulers of this state, who needed (or wanted) money, readily agreed to sell a wide assortment of military equipment including two of the Soviet ZU-23 guns plus ten thousand rounds of ammunition. This was a bonanza and exactly what was needed.

A study was initiated to determine the true capability of the weapon and how it might be counteracted. The study of the ZU-23 quickly produced the answers to our questions about the quality and the capability of the weapon. This twin barreled 23 mm weapon proved to be a light weight, very low profile, gun which was so designed that it could easily be towed by a Jeep-size vehicle then set up and be prepared to fire in forty five seconds. After shutting down it could be prepared for towing and on the road in forty five seconds. This mobility readily explained the apparently high concentration of anti-aircraft guns—one gun could be made to seem like ten because it could be fired from a different location on each pass of the aircraft. Another outstanding feature was the very simple, mechanical tracking sight which required little or no maintenance and allowed the gunner to stay on and track all low level jets except those going directly over the gun.

Those two features alone made this a superior weapon that far exceeded the capability of any weapon we had. One reason was the US decision to put almost all of our anti-aircraft R&D money into missile systems and very little into guns.

We were not able to decide on effective, usable countermeasures. The recommendation was to just keep away from the medium range gun. Not very satisfactory but it helped.

96

AERONAUTICAL COMBAT TEST RANGE

While a member of the Defense Intelligence Agency Scientific & Technical Intelligence team in the Pentagon the author was selected as an intelligence member of the National Range Board. This Board was assigned the responsibility of planning, establishing and overseeing all military test ranges for the United States Government.

The timing included the early stages of the Viet Nam war. This period includes the time of the two earlier chapters when our fighter aircraft losses were deemed unacceptably high during their operation in the extremely hostile environment over North Viet Nam. This is also when it was decided that a training range was needed which would more realistically train pilots for what they would actually encounter when they entered combat. At this time, almost all of the thinking at the planning level was computers, the newest toy. For this reason initial discussions were all based on having the planes fly against a computer simulation of the enemy environment.

As discussions progressed it was suggested that, rather than simulation, we set up an actual enemy environment using the enemy's equipment for our trainees to fly against. A lot of discussion ensued. Some thought that any new operation had to be based on computers or it couldn't work. Others expressed doubt that enough of the enemy equipment could be obtained, operated and maintained in a manner that would reliably assure a technically accurate training program.

Finally the decision was made: flying against the actual equipment would be better training than flying against a simulation, no matter how good the simulation might be. Orders were issued and construction of the range begun. Shortly thereafter, transferred to the test site location the author became the intelligence member of the local range board during its' implementation. Primary responsibility in this phase, was making certain that the installation realistically represented what would be encountered over North Viet Nam.

A large quantity of the desired equipment, including acquisition radars, missile control radars, missile launch units, anti-aircraft guns and other applicable items arrived and the range soon became operational. As the planes approached, the equipment would come on line and do everything except fire the weapons. A computerized scoring systems was installed to inform the pilots, after the fact, whether they won or lost. It quickly became the post-graduate course for fighter pilots headed into combat. It worked. There is also reason to think it saved a lot of lives.

EXPLOSIVE ORDNANCE DISPOSAL

When munitions are manufactured, stored and used there is a need for people trained to handle, defuse and dispose of any item that malfunctions, or otherwise becomes hazardous. Highly trained specialists in each of the military departments, known as Explosive Ordnance Disposal Technicians do this hazardous work. These men, known as EOD Specialists, are also responsible for foreign (enemy!) munitions during hostilities and terrorists activities whenever and wherever they occur.

Primary responsibility for such actions on the ground (anywhere) is assigned to the Army, primary responsibility on the water is assigned to the Navy and the Air Force handles its own airborne munitions wherever needed. Frequently the responsibilities overlap requiring the Services to work together.

During the Viet Nam war the Services were severely tasked. Many booby traps, land mines and other munitions were made in village back yard shops. Some of these were relatively sophisticated while others were very primitive. Since there was no standard production line each weapon was slightly different and had to be handled as a "one of a kind" munition. For this reason each had to be X-rayed and its' fusing mechanism studied before it could be disarmed. One mistake could prove fatal. A lack of suitable X-ray and film development equipment complicated this most serious problem.

In one of the major battle areas of Viet Nam there was one EOD officer to handle all such munitions but he had no equipment. He would gather and stack dangerous items in a pile near a hospital then, when the pile got high enough (100 items when checked), the wing of the hospital containing the X-ray machine would be evacuated so he could carry the items in, one at a time, and X-ray them. Next the film would be sent to another location to be developed. Only then could he study the fusing mechanism and disarm the weapon, learn what he could, then dispose of the remainder. A very primitive operation that was completely unacceptable.

When this situation became known a meeting of approximately thirty of the country's top EOD experts was convened at the Defense Intelligence Agency (DIA). The Army party included Special Forces EOD men. The problem was outlined and suggestions for correcting the situation were solicited from all present. It was learned the Army had only three EOD specialists in Viet Nam at that time. Since the Army had several EOD Companies and also had primary responsibility for land based EOD operations they were asked why none of these Companies was in the only shooting war going on at that time. The Army promised to check into it. (Two weeks later an EOD Company was placed on alert for transfer to Viet Nam.)

The next item discussed was equipment. This was when it was learned the Special Forces had a new X-ray machine which was about one foot cube in size, was designed to be air dropped, could operate off a cigarette lighter in a jeep and had the capability of penetrating up to one inch of steel. This new machine had been developed specifically to meet the needs of the Special Forces and many of the experts present had not even heard of it! Every unit manufactured had been consigned to Special Forces.

Since few of the munitions involved in the problem we were discussing were thicker than one inch this machine seemed ideal for our needs. The possibility of some units being released to other EOD organizations was explored and, about two weeks later, several were made available.

Even though this was a quantum leap ahead there was still one major problem. Even using this highly portable X-ray machine the film still had to get to a photo lab for development before disarming of the weapon could be started. During the discussions representatives of the Naval EOD Laboratory said they had heard rumors the Polaroid Company was working on a field-developed X-ray film process but could not confirm these rumors. It was agreed that a development of this kind might very well solve our problem.

The next week the Polaroid Company was invited to meet for discussion of a serious problem and they responded immediately. Informed of the problem they were asked if, as we had heard, they had something in development which might help. They confirmed they did have such a process under development but advised it was not ready for use. They were asked to accelerate the program and offered development money, if needed. They declined the offer of money but assured us they would immediately step up the pace. They, at our request, also agreed to work with the Navy EOD Laboratory during development and field testing. Within three months the first field developed X-ray film was being tested in Viet Nam. It worked and solved that part of our problem.

Later a large shipment of approximately ten tons of selected live munitions was assembled in Viet Nam and made secure enough for shipment on a transport aircraft to the United States. This shipment satisfied several outstanding requirements without incident.

One of the most interesting facets of this operation was the quality of the people involved. This group of highly qualified specialists were a very close fraternity and most of them already knew each other. When they came together there were no Service lines, no rivalries nor any parochialism shown. There was one hundred per cent cooperation by everyone on every item. The same was true with Polaroid when they joined the team. These were very special men who were totally dedicated to their profession. It was a privilege and a pleasure to meet them and have the opportunity to work with them.

AK-47 ASSAULT WEAPON

In the early days of the Viet Nam war the United States developed two different rapid fire automatic assault weapons for use in combat. These two, the M-15 and the later M-16 greatly increased the firepower of our troops in field combat and gave us a decided advantage.

Shortly thereafter, as often happens, we began to hear of a similar Soviet weapon called the AK-47 and later, a Chinese copy of the Soviet weapon. Both of these began to show up in combat and the US advantage began to disappear. There had been no chance to study and test the Soviet weapon to ascertain what we were up against.

A requirement was levied and collection effort mounted to get one or more of these guns ASAP. Almost immediately we were successful and soon had three of the AK-47 automatic guns in hand.

A meeting of top experts was convened in a vaulted, high security area near the Pentagon. As the meeting time approached the author walked up to the security checkpoint with the three automatic weapons slung over his shoulder. On presenting his credentials the guard denied entry and was obviously perturbed and saying he couldn't permit entry with those guns. All attempts to assure him that everything was in order and that it was necessary to take the guns to the meeting inside were in vain. When he continued to refuse he was asked to contact his supervisor and get clearance to enter because otherwise the meeting couldn't start. He took my name, and other information and started calling but his supervisor and others on up the line wouldn't stick their neck out to allow entry to that secure area with the weapons being carried. Finally it reached the top boss, a three star General. The General told them final authority for approving entry was vested in a specific officer then gave them my name.

When the guard received that information he rechecked my identification. Then he said he had to allow entry but it was quite obvious he was not happy about it.

At the meeting the guns were carefully examined by all the experts present then each of the guns was farmed out to a different organization for test firing. The gun was a very effective weapon and the meeting proved to be very profitable. We learned a lot.

NORTH KOREAN MINIATURE SUBMARINE

During the latter days of the Viet Nam war the author was Commander of a Direct Air Support Center having air support responsibility for the Eastern two-thirds of the Demilitarized Zone (DMZ) between North and South Korea. At the same time he was also Senior Air Advisor to the Commander of the First Republic of Korea (ROK) Army.

One night an eight man North Korean submarine tried to land spies in the area on the coast south of the Demilitarized Zone. The submarine and some of the men were captured. Although no longer working in intelligence the author arranged for the US Navy to have access to the submarine. It was primitive, and of no technical value, but still interesting.

BRUCE STOCKDELL

PRISONERS OF WAR

The latter part of the war in Viet Nam found the author assigned as Chief of Operations for the Deputy Chief of Staff/Intelligence of the 13th Air Force located at Clark Air Force Base in the Philippines. The 13th Air Force had responsibility for all of Southeast Asia, except Viet Nam which was the responsibility of the 7th Air Force. The area included about one fourth of the globe extending to Calcutta, India on the west and half way up China to the north. This was a very large area covering many square miles.

As the war wound down in 1969 a discussion arose as to how Prisoners of War would be handled once they were released. On inquiring we were surprised to learn that there was no plan to meet that eventuality. When the 13th AF Commander, Lt Gen Gideon, was advised of this he directed that a comprehensive plan should be prepared at once and selected the author to be the responsible officer. It was pointed out that the preparation of such a plan would normally fall under the Directorate of Operations while the author was in Intelligence. This was acknowledged, but the General had specified he had picked the man he wanted to do the job. It was an honor to be selected and it turned out to be one the most enjoyable assignments possible in a career.

To assist a Capt Woody—and a T/Sgt Ray—from intelligence were selected.

NOTE; Their last names are not given for they could not be found and their names will not used without their permission. At the same time, however, we want to be certain that they receive full credit for their very excellent work and their valuable contributions to this project.

The task was formidable. First, we had to determine how many prisoners there were; where in Viet Nam or the battle zone they were imprisoned; what service (Army, Air Force, Navy or Marine Corps) they belonged to; where, when and how they they might be released and what would be needed to take care of them. Fortunately the intelligence office had an ongoing program of trying to keep track of Air Force prisoners, their health status and their physical condition. Our information was very limited and very difficult to get but it gave us a place to start.

As we gathered all available information we concluded there were approximately 850 known Prisoners of War and we would have no control over when and where they might be released. Our plan was written to receive and process that number, one at a time, all 850 at one time or any increment in between. The plan also had to be flexible enough to cover release across the battle line in Viet Nam; release from some external international point such as Hong Kong, BCC; Bangkok, Thailand or some other neutral city which could be used by the captors to gain maximum publicity; release in Hanoi, the capital city of North Viet Nam; or release in some other unknown location. Each presented a different set of challenges but the plan had to cover all possibilities.

Our transportation preference was to use USAF C-141 or C-130 aircraft equipped and staffed for medical evacuation but there was no way of knowing whether these aircraft would be permitted to fly into Hanoi if that was the release point chosen. Airlines were considered as a possible alternative but, again we faced possible problems as Air France was the only western airline flying into Hanoi at that time. On contacting Air France, to our relief, they promised to provide up to seven of their planes on forty eight hours notice.

Then we had to decide where to take the men when they were picked up. There was much discussion on this subject. Some people said to the USA ASAP while other, cooler, heads said the cultural shock of jumping from long term imprisonment into full-scale freedom at home, along with the accompanying publicity, might be more than some could cope with. All agreed that a complete, immediate physical checkup, plus a debriefing session, were mandatory. Fortunately there was the perfect place for all these things readily available-Clark Air Force Base in the Philippines. On Clark at that time were several thousand Americans, including entire families with all the amenities it takes to sustain a small American city. A miniature USA if you will. It also had the major USAF hospital that had serviced the war plus a complete, fully staffed field hospital. Both staffs worked the hospital so there was at least two doctors of every known medical speciality. It was the only place known where you could get a routine appointment to have a tooth filled at two-thirty in the morning. It is doubtful that there was a better equipped and staffed hospital anywhere in the world. The decision was made to bring them to Clark for a short "unwinding" period to help them get back on their feet before continuing home.

Next was determining what the men would want or need most upon release. After much discussion the following items were selected for the reasons given:

1. A clean, well fitted uniform-to enable them to shed their prison garb ASAP.

2. A wrist watch-Time would once again mean something to them.

3. A billfold with money in it and a place to spend that money as they wished. This would enable them to buy a hamburger, milkshake, gifts to take home to their loved ones or to use as they wanted. A sum of $1000 per man was proposed.

4. A telephone call home.

5. A chance to completely relax and unwind after their long imprisonment.

Before these items could be written into the 13th Air Force Plan each had to be staffed with the offices that would be responsible. This is where the joy of making this plan came to light. Every office, and and every person contacted was enthusiastic and gave 110% cooperation. There was not one negative in the whole process even though some potentially severe problems were revealed.

When clothing supply offices (Navy for the Navy and Marine Corps and the Air Force for the Air Force and Army) were checked we learned that they did not have enough uniforms in stock to meet the anticipated needs. Both Supply Officers said they would immediately increase their stock holdings so they would not be caught shorthanded.

The Base Exchange Officer said he would increase his stock so there would be a watch assured for every man released.

The Air Force finance officer advised that our projected requirement of $1000 per man for 850 men was nearly double his normal cash reserve but that he would immediately double his reserve so as not to be caught unprepared.

The Base Communications Officer assured us there would be a bank of telephones set up with direct access to the USA.

The Military Airlift Command assured us our air transportation needs, whatever form they took would be met.

Finally, the good people at Clark assured us that each man released would be given the warmest possible welcome. Actually we didn't have to ask that for that is the way the people at Clark were, and we knew it.

This plan was completed, staffed through concerned offices in the Philippines then sent to Pacific Air Forces (PACAF) Hq in Honolulu for approval. About three months passed without response. Then we received, from the Pentagon, a copy of the Air Force Plan for handling and processing any prisoners of war that were released from captivity. It was our plan, word for word. The only changes were the cover and title page. While we at Clark who compiled the plan did not receive even a credit line or honorary mention we did get great satisfaction later in seeing that our plan was apparently followed, to the letter, when the men were released. The quality of our work was very evident when all involved in receiving, processing and caring for these released prisoners, and the prisoners themselves, repeatedly stated that the entire program was superbly planned and executed.

Before this plan became effective there was the first prisoner release. That was the release of twenty two Army men from Cambodia. They were flown directly to Clark where the author, personally, coordinated their arrival-a twenty two hour day. An interesting side issue. Prince Sihanouk of Cambodia had given each man a new uniform tailored by his personal tailor. They were not exactly regulation but they were new and the men looked nice.

Those in Intelligence do get into some unusual, interesting and enjoyable assignments.

B-29 vs CAR

When the author was assigned to fly the B-29 (pp 8) that airplane was still a "secret weapon" as the U.S. public had not yet been told we had such a large, powerful aircraft in our inventory. This plane was also, by far, the most expensive, most complex, aircraft purchased to date at a cost of well over one million dollars (in 1940 dollars) per copy.

Several weeks into this training program the government decided to unveil that beautiful aircraft and to use it to help promote the sale of War bonds. It was a successful and timely move. Until this time his parents could only be told that he was on a highly classified assignment, the nature of which could not be revealed. With the downgrading of the plane's classification they could finally be told he was flying this brand new, huge, very expensive bomber.

Shortly after receiving this information they learned of a bond selling rally to be held in their area featuring a mockup of the front end of the fuselage of a B-29 bomber. The parents made a special effort to see that display. Later they wrote that, on seeing the cockpit of the plane, his mother said "Oh my goodness, just look at that and to think that only two years ago I would not let him have the family car." She was right, Dad would, but she was not in favor. A favorite good memory!!!! It would be interesting to know what her comment would have been if she had known the "Rest of the Story"????

IN MEMORY

of

Capt FRED MOULLEN and Lt MAL STOKES

Three were flying hazardous missions off city

streets behind the IRON CURTAIN.

Two did not return.

BRUCE STOCKDELL

ABOUT THE AUTHOR

The author came from a rural area of southern Indiana during the depression years. His family was not wealthy but his parents were well grounded, solid Christian people who provided well for their family of five children during those difficult years. We never went hungry.

He decided early on that he wanted to be a pilot and an engineer but recognized that, to do so, he would have to leave the farm and make his own way for the family, while encouraging could not financially support his ambitions. Working his way through engineering school at Purdue University he was drafted into the Army in World War II. He immediately asked for Aviation Cadet training but was advised that it was not open and was assigned to basic training in Field Artillery then directly into Field Artillery Officer Candidate School (OCS). Half way through basic training he was offered a chance to apply for Aviation Cadet training and took it.

This change required starting basic training all over again but in the Air Corps. From this point his career changed drastically. It took some very unusual-even bizarre-twists and turns that were totally unpredictable and unexpected. Involved in these changes were adventure, excitement, hazardous escapes, some joy, some sorrow and other unusual experiences that just happened.

While this was going on he acquired a BS in Air Transportation Engineering and a MS in Industrial Administration both from Purdue University. He also finished courses in the German language at the University of Vienna and the Goethe Institute both in Vienna, Austria.

While this is not intended to be an autobiography, the author was present, and personally involved, in every story, every operation, and every example presented in this book. His unusual combination of education and experience and a lifetime attitude of not always knowing all the reasons why an assigned task could not be done but accepting the task and proceeding to get from A to B to best complete the assignment resulted in his being assigned many hard, or even "impossible", task. It also resulted in the last third of his career working almost exclusively at starting up new works for the Air Force. It produced a very challenging, and sometimes fascinating, career.